Advances in Experimental Medicine and Biology

Neuroscience and Respiration

Volume 910

More information about this series at http://www.springer.com/series/13457

Mieczyslaw Pokorski
Editor

Respiratory Medicine and Science

Springer

Editor
Mieczyslaw Pokorski
Public Higher Medical Professional School in Opole
Institute of Nursing
Opole, Poland

ISSN 0065-2598 ISSN 2214-8019 (electronic)
Advances in Experimental Medicine and Biology
ISBN 978-3-319-30658-2 ISBN 978-3-319-30659-9 (eBook)
DOI 10.1007/978-3-319-30659-9

Library of Congress Control Number: 2016940516

Printed on acid-free paper

This Springer imprint is published by Springer Nature
The registered company is Springer International Publishing AG Switzerland

Preface

The book series *Neuroscience and Respiration* presents contributions by expert researchers and clinicians in the field of pulmonary disorders. The chapters provide timely overviews of contentious issues or recent advances in the diagnosis, classification, and treatment of the entire range of pulmonary disorders, both acute and chronic. The texts are thought as a merger of basic and clinical research dealing with respiratory medicine, neural and chemical regulation of respiration, and the interactive relationship between respiration and other neurobiological systems such as cardiovascular function or the mind-to-body connection. The authors focus on the leading-edge therapeutic concepts, methodologies, and innovative treatments. Pharmacotherapy is always in the focus of respiratory research. The action and pharmacology of existing drugs and the development and evaluation of new agents are the heady area of research. Practical, data-driven options to manage patients will be considered. New research is presented regarding older drugs, performed from a modern perspective or from a different pharmacotherapeutic angle. The introduction of new drugs and treatment approaches in both adults and children also is discussed.

Lung ventilation is ultimately driven by the brain. However, neuropsychological aspects of respiratory disorders are still mostly a matter of conjecture. After decades of misunderstanding and neglect, emotions have been rediscovered as a powerful modifier or even the probable cause of various somatic disorders. Today, the link between stress and respiratory health is undeniable. Scientists accept a powerful psychological connection that can directly affect our quality of life and health span. Psychological approaches, by decreasing stress, can play a major role in the development and therapy of respiratory diseases.

Neuromolecular aspects relating to gene polymorphism and epigenesis, involving both heritable changes in the nucleotide sequence and functionally relevant changes to the genome that do not involve a change in the nucleotide sequence, leading to respiratory disorders will also be tackled. Clinical advances stemming from molecular and biochemical research are but possible if the research findings are translated into diagnostic tools, therapeutic procedures, and education, effectively reaching physicians and patients. All that cannot be achieved without a multidisciplinary, collaborative, bench-to-bedside approach involving both researchers and clinicians.

The societal and economic burden of respiratory ailments has been on the rise worldwide leading to disabilities and shortening of life span. COPD alone causes more than three million deaths globally each year. Concerted efforts are required to improve this situation, and part of those efforts are gaining insights into the underlying mechanisms of disease and staying abreast with the latest developments in diagnosis and treatment regimens. It is hoped that the books published in this series will assume a leading role in the field of respiratory medicine and research and will become a source of reference and inspiration for future research ideas.

I would like to express my deep gratitude to Mr. Martijn Roelandse and Ms. Tanja Koppejan from Springer's Life Sciences Department for their genuine interest in making this scientific endeavor come through and in the expert management of the production of this novel book series.

Opole, Poland Mieczyslaw Pokorski

Contents

Advs Exp. Medicine, Biology - Neuroscience and Respiration (2016) 21: 1–8
DOI 10.1007/5584_2015_200

Published online: 29 January 2016

Molecular Characteristics of Influenza Virus Type B Lineages Circulating in Poland

K. Bednarska, E. Hallmann-Szelińska, K. Kondratiuk, D. Rabczenko, and L.B. Brydak

Abstract

From the time of the Hong Kong pandemic of 1968–1969, vaccines against influenza are trivalent, containing two subtypes of influenza type A: A/H1N1/ and A/H3N2/, and influenza type B. In 1980, circulation of the new Yamagata and Victoria lineages of influenza B virus was noted. Since both lineages have continued to circulate, the second lineage of influenza B was included into the trivalent vaccine as of the 2013/2014 epidemic season. In Poland, co-circulation of influenza type A and B has been registered over many seasons, although type A has predominated. According to the ACIP recommendations, quadrivalent vaccines against influenza are administered in some continents due to circulation of the B-Yamagata and B-Victoria lineages. Currently, only trivalent vaccines against influenza are available in Poland. The aim of the present research was to determine which of the two influenza type B lineages, or possibly both, would be isolated in Poland. The study was conducted with the use of RT-PCR. Generally, in the 2014/2015 epidemic season in Poland, circulation of type B virus was confirmed in 34 % of influenza cases. A total of 89 specimens of influenza B were tested, including co-infections of influenza B with influenza A subtypes: A/H1N1/pdm09 and A/H3N2/. The findings were that only lineage B-Yamagata circulates in the Polish population. Therefore, vaccines available on the Polish market do not require the introduction of a fourth component.

K. Bednarska (✉), E. Hallmann-Szelińska, K. Kondratiuk, and L.B. Brydak
Department of Influenza Research, National Influenza Center, National Institute of Public Health – National Institute of Hygiene, 24 Chocimska St., 00-791 Warsaw, Poland
e-mail: karolina.bed88@gmail.com

D. Rabczenko
Centre of Monitoring and Analyses of Population Health Status, National Influenza Center, National Institute of Public Health – National Institute of Hygiene, 24 Chocimska St., 00-791 Warsaw, Poland

Keywords
Epidemic season • Influenza vaccine • Influenza virus • Lineages • Quadrivalent vaccine • Trivalent vaccine

1 Introduction

Discovery of influenza virus by Christopher Andrewes, Patrick Laidlaw, and Wilson Smith took place in 1933 in the Medical Research Council's National Institute for Medical Research in London. The Worldwide Institute for Influenza designated by the World Health Organization has been located in aforementioned institute. Influenza in humans can be caused by influenza viruses type A and type B. Despite the fact that influenza A is responsible for most of seasonal infections, influenza type B is common in children and young adults. Seasonal influenza epidemics result in considerable rates of morbidity and mortality. Therefore, public health agencies around the world recommend vaccination against influenza to protect population from seasonal epidemics (Li et al. 2008; Hite et al. 2007; Olson et al. 2007). Trivalent vaccines against influenza consist of A/H1N1/ and A/H3N2/, the two subtypes of influenza type A, and influenza type B.

The Yamagata lineage had been the main circulating lineage of influenza type B until 1980 when a second antigenically different B/Victoria lineage appeared. Since that time on both lineages have been co-circulating globally (McCullers et al. 2004; Rota et al. 1990). There is a low or no cross-reactive protection between B/Yamagata and B/Victoria lineages, which means that appropriate protection from influenza B depends on the correctly anticipated dominant influenza B lineage in every forthcoming season (Ambrose and Levin 2012; Belshe 2010). In the 2013/2014 season, due to continuing circulation of both lineages of influenza virus B in populations of some regions of the world, the second lineage of influenza B was included into the trivalent vaccine against influenza (WHO 2013).

The aim of the present research was to determine which of the two influenza type B lineages, or possibly both, would circulate in Poland. Currently, only trivalent vaccines against influenza are available in Poland. Therefore, the exact knowledge on the circulating lineages would be essential to acquire the accordingly modified vaccines.

2 Methods

The study protocol was approved by an institutional Ethics Committee and the study was conducted in accordance with the principles for biomedical human research as set by the Declaration of Helsinki.

2.1 Patient Population and Specimen Collection

There were 89 clinical specimens, positive for influenza virus B, collected from all over Poland, tested at the Department of Influenza Research of the National Influenza Center of the National Institute of Public Health – National Institute of Hygiene in Warsaw, Poland. Specimens consisted of nasal or throat swabs collected during the 2014/2015 epidemic season. All specimens and virus isolates were stored at −80 °C prior to procedures.

2.2 Extraction of Viral RNA

The viral RNA was extracted from 200 μl of clinical samples in viral transport medium consisting of phosphate-buffered saline, using a Maxwell 16 Viral Total Nucleic Acid Purification Kit according to the manufacturer's instructions for Low Elution Volume (LEV) cartridges (Promega Corporation; Madison, WI). The RNA was eluted with 50 μl of RNase-free water.

2.3 Reverse Transcription Polymerase Chain Reaction (RT-PCR)

The detection of influenza subtypes was performed by one-step RT-PCR reaction using Roche Light Cycler 2.0 System (Roche Diagnostics, Rotkreuz, Switzerland). RT-PCR reactions were performed in capillary tubes of 20 μl volume with 0.5 μl (20 nM) primers and 0.5 μl (5 nM) probes for each reaction. Primers and probes were obtained through the Influenza Reagent Resource (IRR) program of the American Centers for Disease Prevention and Control. The reaction mixture, containing reaction buffer, $MgSO_4$ buffer, bovine serum albumin (BSA), RNase-free H_2O, and SuperScript® III/Platinum® *Taq* Mix (Invitrogen by Life Technologies – Thermo Fisher Scientific, Carlsband, CA) was incubated with 5 μl RNA sample per capillary tube. RNA from viruses B/Massachusetts/2/2012 (B/Yamagata) and B/Brisbane/60/2008 (B/Victoria) were introduced as positive controls and RNase-free H_2O was utilized as a negative control sample. Before DNA amplification cycles were begun, RNA templates were reverse transcribed to produce the corresponding cDNA templates during reverse transcription procedure: 50 °C for 30 min. DNA templates were then subjected to an initialization step (1 cycle at 95 °C for 2 min), followed by 45 cycles of amplification: denaturation at 95 °C for 15 s, annealing at 55 °C for 30 s, and elongation at 72 °C for 20 s.

2.4 Cell Culture and Virus Passaging

MDCK cells (ATCC® CCL-34™; Teddington, UK) were cultivated in Dulbecco's Modified Eagle Medium (DMEM) with antibiotics (penicillin 1000 U/ml with streptomycin 1 mg/ml) and fetal bovine serum to a final concentration of 10 % into 5 cm^2 tubes and allowed to grow to confluence at 37 °C in 5 % CO_2. Monolayers were washed twice with Ca^{2+}/Mg^{2+} -free phosphate-buffered saline before incubation with 50 μl virus sample (clinical isolates were inoculated neat; cell- or egg-passaged virus was inoculated at 1/100 dilution) at room temperature for 30 min. After inoculation, 3 ml of fetal bovine serum-free medium, supplemented with TPCK trypsin to a final dilution of 2.5 μg/ml, was added to each tube and the cells were incubated at 35 °C and 5 % CO_2. The wells were monitored daily for virus growth by cytopathic effect (CPE). The supernatant was collected after 4 days and the presence of virus was assessed by a hemagglutination assay (HA). The assay was performed using 0.75 % turkey red blood cells in phosphate buffer saline (PBS) of pH 7.2 in V-bottom 96-well plated (Nunc #651101) by a standard technique (WHO 2011).

2.5 Antigenic Affinity (Hemagglutinin Inhibition Assay) and Sequencing of B/Yamagata Isolates

Within global influenza surveillance, the National Influenza Centre was obliged to collect and forward viral specimens to a WHO Collaborating Centre where further characteristics of strains circulating in Poland were investigated. Specifically, antigenic affinity to different strains of influenza viruses was determined and sequencing was performed there. This kind of analysis was necessary to optimize the composition of a vaccine against influenza for the forthcoming season. The following three B/Yamagata isolates were tested: B/Poland/2477/2015 Mar, B/Poland/1062/2015 Feb, B/Poland/ 1054/2015 Mar.

3 Results

The outcome of RT-PCR clearly indicates that B/Yamagata viruses only circulated in the

2013/2014 epidemic season in Poland. All tested strains of influenza viruses belonged to this lineage. At this point, B/Victoria lineage was not confirmed in any specimen.

All influenza B viruses were successfully propagated in the MDCK cell line. The fluids harvested were tested by HA to estimate the viral titer. It ranged from 128 to 1024 HAU (hemagglutination units). Further, antigenic affinity, as assessed by hemagglutinin inhibition assay, demonstrates that the three isolates tested were recognized by the antiserum raised against the egg-propagated cultivar of the vaccine virus B/Phuket/3073/2013 – at titers only two-fold reduced or above the titer for the homologous virus. The antiserum raised against the cell culture-propagated cultivar of B/Phuket/3073 recognized two test viruses at titers equal to the titer of the antiserum for the homologous virus and one test virus – B/Poland/1054/2015, at a titer within 4-fold of the titer of the antiserum for the homologous virus. Two of the three test viruses were recognized at titers within 4-fold of the titer for the homologous virus by the antiserum raised against the previously recommended egg-propagated vaccine virus B/Massachusetts/02/2012. All three test viruses were recognized at titers within 4-fold of the titer for the homologous virus by the antiserum raised against the cell culture-propagated cultivar of B/Massachusetts/02/2012, two at an equal titer to the homologous titer (Table 1).

Additionally, sequence analysis of the hemagglutinin (HA) and neuraminidase (NA) genes on all three of the isolates was carried out. All clustered in genetic clade 3, the B/Wisconsin/1/2010 – B/Phuket/3073/2013 clade (Figs. 1 and 2). Clade 3 viruses predominated globally over more than the last 12 months.

4 Discussion

There are intense discussions worldwide regarding a global implementation of the quadrivalent (QIV) vaccine against influenza, which would contain two lineages of influenza B virus. Currently, most European countries have a trivalent (TIV) vaccine composed of two strains of influenza type A, A/H1N1/pdm09 and A/H3N2/, and one strain of influenza type B virus. Initially, QIV was introduced in the epidemic season of 2013/2014. Findings in the literature suggest that inclusion of both lineages of influenza B virus into the QIV would have a positive influence in terms of reducing hospitalization related to influenza infection and would be cost saving. It also would prevent influenza infections in a more efficient way (You et al. 2014; Lee et al. 2012; Reed et al. 2012). More importantly, inclusion of the fourth component in the vaccine does not adversely affect the safety or immunogenicity of QIV compared with TIV vaccine. The only difference in the manufacturing process between both types of vaccine is an addition of the second lineage of influenza B strain in the final formulation (Greenberg et al. 2013). Another concern about the quadrivalent vaccine has been that its production would fall short of increasing demand for it. It seems now that the manufacturing industry is capable of meeting the demand for QIV (Ambrose and Levin 2012; Ambrose and Levin 2015).

Predomination of influenza B has not been registered in recent epidemic seasons in Poland. However, some increase in circulation of influenza type B has been noted during the season of 2014/2015. The circulating strain belonged to the B/Yamagata lineage and it was related to the vaccine strain recommended for that season, namely B/Massachusetts/2/2012. A phylogenetic comparison of HA and NA demonstrates that the Polish isolates belonged to clade 3 that contains the strain B/Phuket (B/Yamagata lineage) included in the vaccine against influenza for the forthcoming 2015/2016 season.

Although the present study shows that the B/Yamagata lineage was the one circulating in Poland in the last epidemic season, the presence of the B/Victoria lineage cannot be firmly excluded since just a small percentage of circulating viruses is actually tested within the framework of influenza surveillance. Moreover, a high variability of influenza virus may lead to the appearance of a second lineage among the viruses circulating in Poland. The Polish

Table 1 Results of hemagglutinin inhibition assay of Polish isolates of influenza B viruses

Hemagglutination inhibition titer								
Post infection ferret antisera								
				B/Phucket	B/Mass	B/Mass	B/Phucket	B/Phucket
				3073/13	02/12	0/12	3073/13	3073/13
		Collection date	Passage history	SH614	T/C F42/14	F15/13	Egg F36/14	T/C F35/14
	Genetic group			1	2	2	3	3
Reference viruses								
B/Florida/4/2006	1	2006-12-15	E7/E1	1280	640	160	160	20
B/Brisbane/3/2007	2	2007-09-03	E2/E3	1280	640	80	160	10
B/Wisconsin/1/2010	3	2010-02-20	E3/E3	2560	320	40	160	40
B/Stockholm/12/2011	3	2011-03-28	E4/E1	1280	160	40	80	20
B/Estonia/55669/2011	2	2011-03-14	MDCK2/MDCK3	1280	160	320	160	40
B/Massachusetts/02/2012	2	2012-03-13	E3/E3	1280	1280	160	160	10
B/Massachusetts/02/2012	2	2012-03-13	MDCK1/C2/MDCK3	1280	640	320	160	40
B/Phuket/3073/2013	3	2013-11-21	E4/E3	2560	320	40	320	40
B/Phuket/3073/2013	3	2013-11-21	MDCK2/MDCK2	5120	320	320	640	640
B/Hong Kong/3417/2014	3	2014-06-04	E4/E1	1280	80	40	80	10
Test viruses								
B/Poland/1054/2015	3	2015-03-27	MDCK1/MDCK1	2560	160	80	160	160
B/Poland/2477/2015	3	2015-03-16	MDCK1/MDCK1	5120	320	320	640	640
B/Poland/1062/2015	3	2015-02-26	MDCK1/MDCK1	5120	640	320	640	640

Based on data from Worldwide Influenza Center, Francis Crick Institute, London 2015

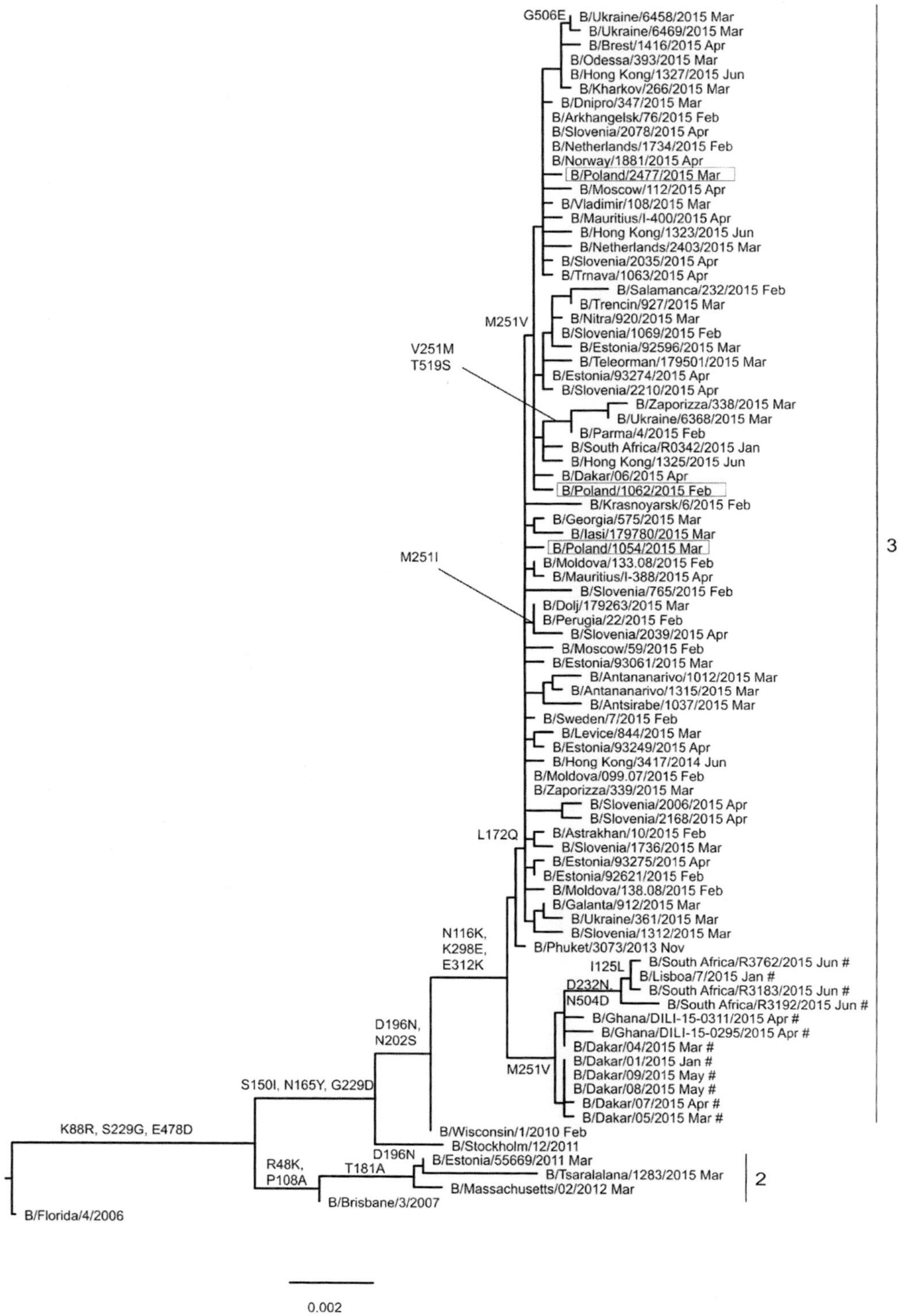

Fig. 1 Phylogenetic comparison of influenza B (Yamagata lineage) HA genes performed by the Worldwide Influenza Center, Francis Crick Institute in London. Polish isolates are in *brackets*

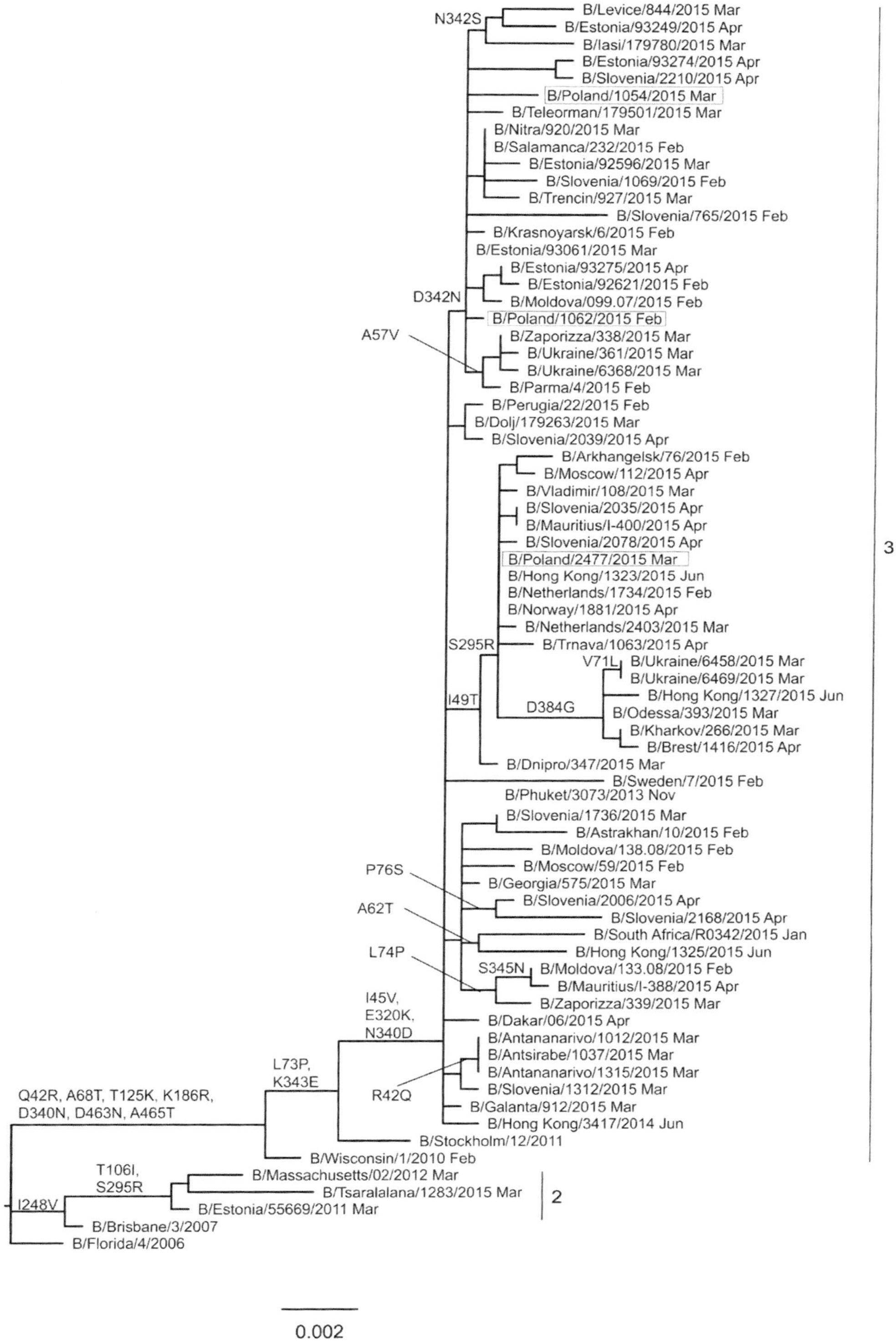

Fig. 2 Phylogenetic comparison of influenza B (Yamagata lineage) NA genes performed by the Worldwide Influenza Center, Francis Crick Institute in London. Polish isolates are in *brackets*

National Influenza Centre will continue to monitor and analyze the circulating strains of influenza type B to evaluate a potential need arising for the quadrivalent vaccine.

Even if there is an epidemic season in which influenza B circulates at a minimal degree, vaccination with the QIV would still be beneficial since it increases immunity to both lineages of influenza B in subsequent seasons. In addition, people traveling to countries where the other lineage of influenza B prevails would be protected. From the standpoint of public health, if the QIV vaccine leads to fewer mismatched seasonal vaccine campaigns, people would be more willing to accept vaccination. That is essential particularly in case of Poland where the current ratio of vaccination equals 3.55 % of population.

Acknowledgements This study was funded in parts by grants 2011/01/B/NZ7/06188 and NIPH-NIH's subject 5/EM.1. We wish to acknowledge the team from the Worldwide Influenza Center of the Crick Institute in London, headed by John McCauley, for the analysis of Polish isolates within the Global Influenza Surveillance and Response System. We also thank the physicians and employees of VSESs participating in the SENTINEL program for their input into the influenza surveillance in Poland.

Conflicts of Interest The authors declare no conflicts of interest in relation to this article.

References

Ambrose CS, Levin MJ (2012) The rationale for quadrivalent influenza vaccines. Hum Vaccine Immunother 8(1):81–88

Ambrose CS, Levin MJ (2015) Quadrivalent influenza vaccines (QIV) – what are they, safety, cost, benefit information. Vaccine; pii: S0264-410X(15)01416-4

Belshe RB (2010) The need for quadrivalent vaccine against seasonal influenza vaccine. Vaccine 288: D45–D53

Greenberg DP, Robertson CA, Noss MJ, Blatter MM, Biedenbender R, Decker MD (2013) Safety and immunogenicity of quadrivalent inactivated influenza vaccine compared to licensed trivalent inactivated influenza vaccines in adults. Vaccine 31:770–776

Hite LK, Glezen WP, Demmler GJ, Munoz FM (2007) Medically attended pediatric influenza during the resurgence of the Victoria lineage of influenza B virus. Int J Infect Dis 11:40–47

Lee BY, Bartsch SM, Willig AM (2012) The economic value of a quadrivalent versus trivalent influenza vaccine. Vaccine 30:7443–7446

Li WC, Shih SR, Huang YC, Chen GW, Chang SC, Hsiao MJ, Tsao KC, Lin TY (2008) Clinical and genetic characterization of severe influenza B-associated diseases during an outbreak in Taiwan. J Clin Virol 42:45–51

McCullers A, Saito T, Iverson AR (2004) Multiple genotypes of influenza B virus circulated between 1979 and 2003. J Virol 78:12817–12828

Olson DR, Heffernan RT, Paladini M, Konty K, Weiss D, Mostashari F (2007) Monitoring the impact of influenza by age: emergency department fever and respiratory complaint surveillance in New York City. PLoS Med 4(8):e247. doi:10.1371/journal.pmed.0040247

Reed C, Meltzer MI, Finelli L, Fiore A (2012) Public health impact of including two lineages of influenza B in a quadrivalent seasonal influenza vaccine. Vaccine 30:1993–1998

Rota PA, Wallis TR, Harmon MW, Rota JS, Kendal AP, Nerome K (1990) Cocirculation of two distinct evolutionary lineages of influenza type B virus since 1983. Virology 175:59–68

WHO – World Health Organization (2011) Global influenza surveillance network. Manual for the laboratory diagnosis and virological surveillance of influenza. World Health Organization, Geneva, 140 p

WHO – World Health Organization (2013) Recommended composition of influenza virus vaccines for use in the 2013–2014 northern hemisphere influenza season. Wkly Epidemiol Rec 88:101–116

You JH, Ming WK, Chan PK (2014) Cost–effectiveness analysis of quadrivalent influenza vaccine versus trivalent influenza vaccine for elderly in Hong Kong. BMC Infect Dis 14:618. doi:10.1186/s12879-014-0618-9

Advs Exp. Medicine, Biology - Neuroscience and Respiration (2016) 21: 9–14
DOI 10.1007/5584_2016_215

Published online: 18 March 2016

Obstructive Sleep Apnea Is Related to Increased Arterial Stiffness in Ultrasound Speckle-Tracking Analysis

I. Tuleta, D. Skowasch, J. Krycki, C. Pizarro, C. Hammerstingl, M. Weber, N. Schahab, G. Nickenig, C. Schaefer, and S. Pingel

Abstract

Obstructive sleep apnea (OSA) is an independent risk factor for atherosclerosis. The aim of our study was to determine arterial stiffness in OSA patients by means of the ultrasound speckle-tracking-based method. Twenty six OSA patients and 17 control subjects were enrolled in the study. The speckle-tracking-based analysis of carotid artery included circumferential strains, circumferential strain rates, radial displacement, and radial strain rates. We found that the global average circumferential strains, circumferential strain rates, and radial displacement were significantly lower in OSA patients compared to controls (2.19 ± 0.30 % *vs.* 4.17 ± 0.33 %, 0.22 ± 0.03 l/s *vs.* 0.31 ± 0.02 l/s, 0.10 ± 0.01 mm *vs.* 0.16 ± 0.02 mm, respectively, $p < 0.05$ for all). There were no significant differences in radial strain rates between the groups (0.32 ± 0.04 % *vs.* 0.33 ± 0.01 %). We conclude that OSA is associated with an increased arterial stiffness.

Keywords

Functional vessel damage • Intermittent nocturnal hypoxia • Vascular incompliance • Vessel wall deformation • Vessel wall motion

1 Introduction

OSA is an established cardiovascular risk factor (Marshall et al. 2008) which may impair cardiovascular status independently of other OSA-associated risk parameters such as arterial hypertension, diabetes mellitus, or hypercholesterolemia. The mechanisms underlying OSA-related cardiovascular worsening are

I. Tuleta (✉), D. Skowasch, J. Krycki, C. Pizarro, C. Hammerstingl, M. Weber, N. Schahab, G. Nickenig, C. Schaefer, and S. Pingel
Department of Internal Medicine II – Cardiology, Pulmonology and Angiology, University of Bonn, 25 Sigmund-Freud-St., D-53105 Bonn, Germany
e-mail: Izabela.Tuleta@ukb.uni-bonn.de

complex and involve stimulation of sympathetic nerves, imbalance between vasoconstrictors and vasorelaxants, elevated oxidative stress, inflammation, endothelial dysfunction, decreased endothelial regeneration capacity, fluctuations of intrathoracic pressure, hypercoagulability, and metabolic changes (Tuleta et al. 2011). A number of studies have reported on the increase in the atherosclerotic plaque load in OSA patients. The arterial stiffness reflecting a subclinical stage of atherosclerosis is also elevated in OSA individuals (Herrington et al. 2004). In previous studies, arterial stiffness in OSA patients has been assessed by means of different parameters such as the central pulse wave velocity (Schaefer et al. 2015; Drager et al. 2005), the augmentation index (Kohler et al. 2008), and the cardio-ankle vascular index (Kumagai et al. 2009), which are sensitive parameters of incipient vessel changes. The aim of the present study was to examine early arterial functional alternations in OSA patients by the ultrasound speckle-tracking technique which is a relatively new method to determine arterial stiffness (Saito et al. 2012; Bjällmark et al. 2010).

2 Methods

2.1 Patients

The present study was conducted according to the principles of the Declaration of Helsinki for Human Research and was approved by the local ethics committees. Written informed consent was obtained from all subjects. Our study was conducted in the pulmonary division of the University of Bonn between May and September 2015. Obstructive sleep apnea was confirmed polysomonographically in 26 patients. Apnea was defined as an air flow cessation for at least 10 s, hypopnea was characterized by air flow reduction of more than 50 % from the baseline for at least 10 s. Apnea-hypopnea-index was determined as the sum of apnea-hypopnea events per hour of sleep. Seven patients suffered from mild OSA ($5 < AHI \leq 15$/h) and the remaining 19 patients from moderate-to-severe OSA (AHI >15/h). Seventeen control patients were free of OSA in the cardiorespiratory polygraphy analysis (AHI <5). The OSA patients had significantly higher body mass index (BMI) compared to controls. There were no otherwise relevant differences in the baseline characteristics between the OSA and control subjects; especially the classical, other than OSA, cardiovascular risk factors were distributed similarly in both groups (Table 1).

2.2 Angiological Examinations

Angiological measurements in the OSA patients were performed following polysomnography, prior to the introduction of CPAP treatment. The study subjects were examined in the supine position. The common carotid artery, one centimeter inferior to the carotid bulb was shown in the two-dimensional short axis in cardiovascular ultrasound echocardiography (Philips iE33, Hamburg, Germany). Thereafter, at least four electrocardiography (ECG)-triggered cardiac

Table 1 Patients' baseline characteristics

	Age (years)	Male gender	BMI (kg/m^2)	Arterial hypertension	Diabetes mellitus	Nicotine abuse			Hypercholesterolemia
						Never	Former	Current	
OSA	60.1	20/26	32.6	20/26	5/26	8/21	10/21	3/21	7/26
(n = 26)	±2.5	(76.9 %)	±1.0	(76.9 %)	(19.2 %)	(38.1 %)	(47.6 %)	(14.3 %)	(26.9 %)
Control	57.6	9/17	23.2	10/16	0/17	9/17	5/17	3/17	4/14
(n = 17)	±1.4	(52.9 %)	±0.7*	(62.5 %)	(0.0 %)	(52.9 %)	(29.4 %)	(17.6 %)	(28.6 %)

Data are means ±SE or n (%)
OSA obstructive sleep apnea, *BMI* body mass index
*$p < 0.05$

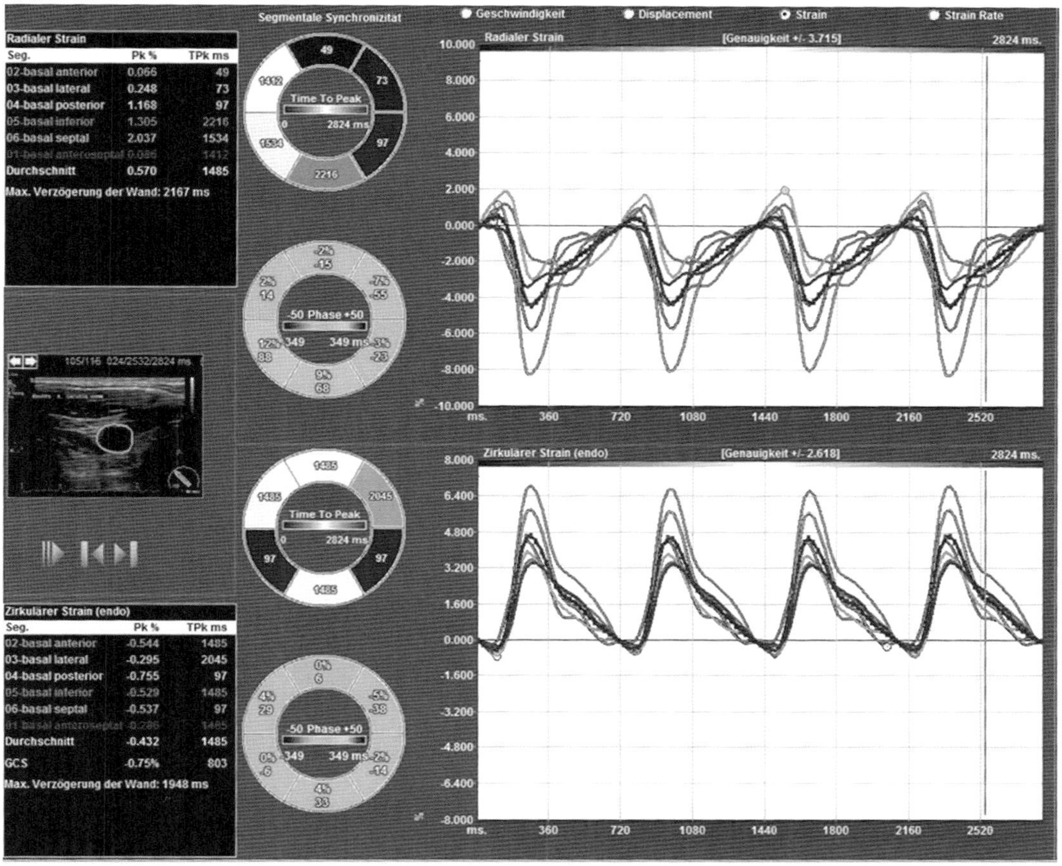

Fig. 1 Typical strain curves of four cardiac cycles of common carotid artery from an OSA patient

cycles were acquired. The obtained images were exported to a computer workstation equipped with a speckle-tracking software package for an off-line analysis (Image Arena Version 4.6, TomTec Systems GmbH, Munich, Germany). Speckle-tracking curves, as exemplified in Fig. 1, were obtained in the region of interest. The strains were determined by tracking interference patterns across the imaging frames. The analysis included the global average circumferential strains (% of vessel circumference changes), global average circumferential strain rates (vessel circumference changes per second), radial displacement (changes in vessel wall position in mm), and radial strain rates (% of vessel wall thickness changes per second), as previously described (Larsson et al. 2014; Yang et al. 2013).

2.3 Statistical Analysis

Differences in mean values between both groups were assessed by Student's *t*-test. Categorical variables were compared by the Chi-squared test. A p-value of <0.05 was considered as statistically significant. In order to exclude the influence of potential confounding factors on the vascular status, the general linear model with a covariate variable was used. The statistical analysis was performed using a statistical SPSS package, version 22.

3 Results

The ultrasonographic analysis of the vessel motion showed that the mean of circumferential

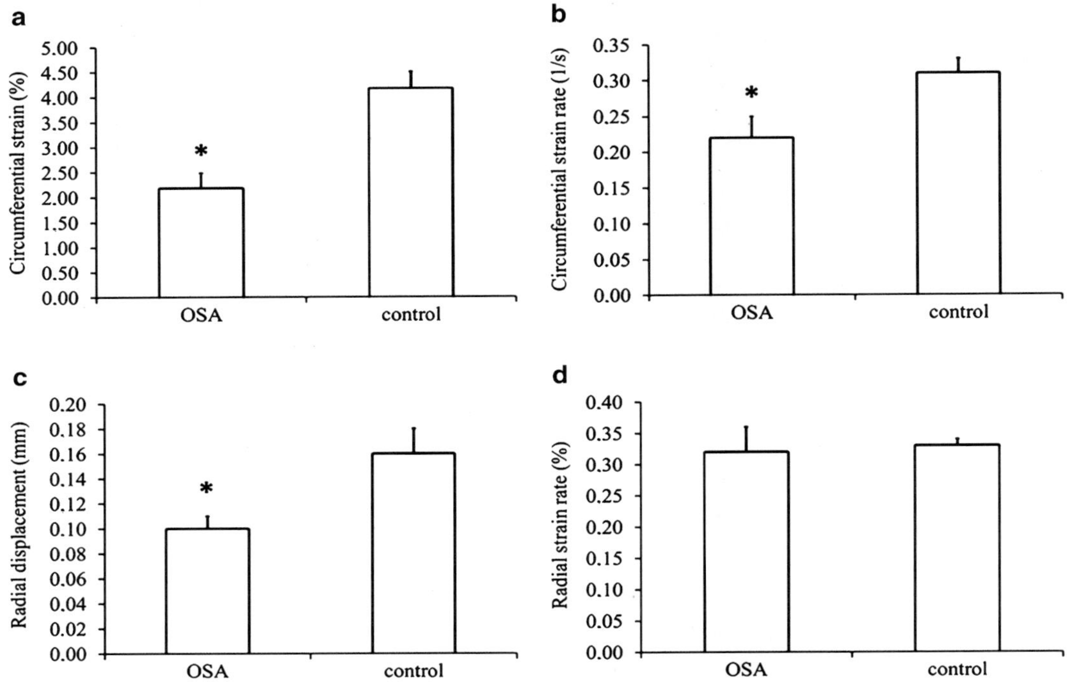

Fig. 2 Vascular strain imaging of the common carotid artery in OSA patients. The mean values of circumferential strain (**a**), circumferential strain rate (**b**), and radial displacement (**c**) were significantly reduced in OSA patients *vs.* controls. No relevant differences in the radial strain rate between the OSA and control subjects were found (**d**); *$p < 0.05$

strain values for all 6 carotid artery segments was significantly reduced in the OSA patients compared to the control group (2.19 ± 0.30 % *vs.* 4.17 ± 0.33 %, respectively, $p < 0.05$) (Fig. 2a). Analogically, the mean values of circumferential strain rates and radial displacement were relevantly lower in the OSA *vs.* control subjects (0.22 ± 0.03 1/s *vs.* 0.31 ± 0.02 1/s; 0.10 ± 0.01 mm *vs.* 0.16 ± 0.02 mm, respectively, $p < 0.05$) (Fig. 2b, c). In contrast, there were no differences in the radial strain rates between the groups (0.32 ± 0.04 % *vs.* 0.33 ± 0.01 %, respectively, Fig. 2d). The analysis of the OSA subgroups demonstrated a trend toward a greater impairment of vascular function in OSA with AHI ≥15 *vs.* OSA with AHI <15 (circumferential strain: 2.07 ± 0.34 % *vs.* 2.50 ± 0.64 %, circumferential strain rate: 0.21 ± 0.04 1/s *vs.* 0.26 ± 0.07 1/s, and radial displacement: 0.09 ± 0.01 mm *vs.* 0.11 ± 0.04 mm, respectively). However, statistical significance could be achieved only while comparing single subgroups to control and not between both subgroups. Since BMI was the only baseline parameter which differed significantly between the OSA and control groups, we could exclude its influence on the circumferential strains or radial displacement in the general linear model with BMI as a covariate.

4 Discussion

Our data demonstrate a significantly reduced vascular function in OSA patients using a novel technique of vascular strain imaging. This technique has already been applied to evaluate cardiac function in OSA patients (Wang et al. 2015). Our study is one of the first reports on alternations in vascular compliance by means of this method in OSA individuals. Although strain values correlate significantly with conventional indexes of vascular wall changes (Kawasaki et al. 2009), circumferential strains seem to be more sensitive in the detection of age-related decreases in artery elastic properties than the classical measures of arterial stiffness (Bjällmark et al. 2010). Additionally, circumferential strains are stronger correlated to the severity of coronary atherosclerosis than a well established surrogate marker for atherosclerosis such as intima-media thickness (IMT) (Kim et al. 2012). Moreover, changes in arterial strain values may be helpful in the determination of plaque vulnerability (Liang et al. 2009) and cardiovascular risk (Catalano et al. 2011). Reduced strain values underlying increased arterial stiffness have been demonstrated in other disorders such as arterial hypertension, diabetes mellitus, renal diseases, Marfan syndrome, and Takayasu arteritis (Zhang et al. 2014; Park et al. 2013; Yang et al. 2013; Cho et al. 2010; Yang et al. 2010). In our study, in contrast to the results for circumferential strains and circumferential strain rates, there were no differences in the radial strain rates between the OSA and normal groups. That may be explained by a lower sensitivity for arterial changes of the radial strains due to a higher variability of single radial strain variables (Bjällmark et al. 2010). Since the OSA patients of the present study were obese compared to controls, we have excluded the potential influence of BMI as a confounding factor on the circumferential strains and radial displacement in the statistical analysis and obtained still significant results. However, BMI may have an impact on the arterial stiffness *via* its influence on the AHI score (Iguchi et al. 2013). In the present study, OSA patients with the lower AHI of less than 15/h showed a tendency toward higher strain parameters compared to those with the AHI of more than 15. Similar results have been obtained for other OSA populations in whom pulse wave velocity correlated significantly with AHI (Drager et al. 2005), which suggests an association between OSA severity and arterial stiffness.

In conclusion, ultrasound analysis of arterial motion and deformation demonstrates a higher arterial stiffness in OSA patients compared to healthy subjects. This finding confirms that OSA is an independent risk factor for impaired mechanical properties of arteries which mirror a preclinical stage of atherosclerosis. Due to the application of a speckle-tracking method, OSA patients at high risk of atherosclerosis development, despite absence of any symptoms of vessel disease and/or atherosclerotic plaques, may be early identified and adequately treated to prevent future cardiovascular complications.

Acknowledgments We gratefully acknowledge Sonotechnik Austria for providing us with the AngE Pro8®. We would also like to thank Dr. R. Fimmers, Institute of Medical Biometry, Informatics and Epidemiology, University of Bonn, for his statistical help.

Conflicts of Interest The authors declare no conflicts of interest in relation to this article.

References

Bjällmark A, Lind B, Peolsson M, Shahgaldi K, Brodin LÅ, Nowak J (2010) Ultrasonographic strain imaging is superior to conventional non-invasive measures of vascular stiffness in the detection of age-dependent differences in the mechanical properties of the common carotid artery. Eur J Echocardiogr 11:630–636

Catalano M, Lamberti-Castronuovo A, Catalano A, Filocamo D, Zimbalatti C (2011) Two-dimensional speckle-tracking strain imaging in the assessment of mechanical properties of carotid arteries: feasibility and comparison with conventional markers of subclinical atherosclerosis. Eur J Echocardiogr 12:528–535

Cho IJ, Shim CY, Yang WI, Kim SA, Chang HJ, Jang Y, Chung N, Ha JW (2010) Assessment of mechanical properties of common carotid artery in Takayasu's arteritis using velocity vector imaging. Circ J 74:1465–1470

Drager LF, Bortolotto LA, Lorenzi MC, Figueiredo AC, Krieger EM, Lorenzi-Filho G (2005) Early signs of atherosclerosis in obstructive sleep apnea. Am J Respir Crit Care Med 172:613–618

Herrington DM, Brown WV, Mosca L, Davis W, Eggleston B, Hundley WG, Raines J (2004) Relationship between arterial stiffness and subclinical aortic atherosclerosis. Circulation 110:432–437

Iguchi A, Yamakage H, Tochiya M, Muranaka K, Sasaki Y, Kono S, Shimatsu A, Satoh-Asahara N (2013) Effects of weight reduction therapy on obstructive sleep apnea syndrome and arterial stiffness in patients with obesity and metabolic syndrome. J Atheroscler Thromb 20:807–820

Kawasaki T, Fukuda S, Shimada K, Maeda K, Yoshida K, Sunada H, Inanami H, Tanaka H, Jissho S, Taguchi H,

Yoshiyama M, Yoshikawa J (2009) Direct measurement of wall stiffness for carotid arteries by ultrasound strain imaging. J Am Soc Echocardiogr 22:1389–1395

Kim SA, Park SM, Kim MN, Kim YH, Cho DH, Ahn CM, Hong SJ, Lim DS, Shim WJ (2012) The relationship between mechanical properties of carotid artery and coronary artery disease. Eur Heart J Cardiovasc Imaging 13:568–573

Kohler M, Craig S, Nicoll D, Leeson P, Davies RJ, Stradling JR (2008) Endothelial function and arterial stiffness in minimally symptomatic obstructive sleep apnea. Am J Respir Crit Care Med 178:984–988

Kumagai T, Kasai T, Kato M, Naito R, Maeno K, Kasagi S, Kawana F, Ishiwata S, Narui K (2009) Establishment of the cardio-ankle vascular index in patients with obstructive sleep apnea. Chest 136:779–786

Larsson M, Heyde B, Kremer F, Brodin LÅ, D'hooge J (2014) Ultrasound speckle tracking for radial, longitudinal and circumferential strain estimation of the carotid artery: an in vitro validation via sonomicrometry using clinical and high-frequency ultrasound. Ultrasonics 56:399–408

Liang Y, Zhu H, Friedman MH (2009) The correspondence between coronary arterial wall strain and histology in a porcine model of atherosclerosis. Phys Med Biol 54:5625–5641

Marshall NS, Wong KK, Liu PY, Cullen SR, Knuiman MW, Grunstein RR (2008) Sleep apnea as an independent risk factor for all-cause mortality: the Busselton health study. Sleep 31:1079–1085

Park DW, Kruger GH, Rubin JM, Hamilton J, Gottschalk P, Dodde RE, Shih AJ, Weitzel WF (2013) In vivo vascular wall shear rate and circumferential strain of renal disease patients. Ultrasound Med Biol 39:241–252

Saito M, Okayama H, Inoue K, Yoshii T, Hiasa G, Sumimoto T, Nishimura K, Ogimoto A, Higaki J (2012) Carotid arterial circumferential strain by two-dimensional speckle tracking: a novel parameter of arterial elasticity. Hypertens Res 35:897–902

Schaefer CA, Adam L, Weisser-Thomas J, Pingel S, Vogel G, Klarmann-Schulz U, Nickenig G, Pizarro C, Skowasch D (2015) High prevalence of peripheral arterial disease in patients with obstructive sleep apnoea. Clin Res Cardiol 104:719–726

Tuleta I, Pabst S, Juergens UR, Nickenig G, Skowasch D (2011) Obstructive sleep apnoea as a risk factor for atherosclerosis – implication for preventive and personalised treatment. EPMA J 2:39–47

Wang D, Ma GS, Wang XY, Lu QQ, Wang Y, Liu NF (2015) Left ventricular subclinical dysfunction associated with myocardial deformation changes in obstructive sleep apnea patients estimated by real-time 3D speckle-tracking echocardiography. Sleep Breath May 24 (Epub ahead of print)

Yang WI, Shim CY, Cho IJ, Chang HJ, Choi D, Jang Y, Chung N, Cho SY, Ha JW (2010) Dyssynchronous systolic expansion of carotid artery in patients with Marfan syndrome. J Am Soc Echocardiogr 23:1310–1316

Yang EY, Brunner G, Dokainish H, Hartley CJ, Taffet G, Lakkis N, Taylor AA, Misra A, McCulloch ML, Morrisett JD, Virani SS, Ballantyne CM, Nagueh SF, Nambi V (2013) Application of speckle-tracking in the evaluation of carotid artery function in subjects with hypertension and diabetes. J Am Soc Echocardiogr 26:901–909

Zhang L, Yin JK, Duan YY, Liu X, Xu L, Wang J, Yang YL, Yuan LJ, Cao TS (2014) Evaluation of carotid artery elasticity changes in patients with type 2 diabetes. Cardiovasc Diabetol 13:39. doi:10.1186/1475-2840-13-39

Advs Exp. Medicine, Biology - Neuroscience and Respiration (2016) 21: 15–21
DOI 10.1007/5584_2015_202

Published online: 29 January 2016

Sclerostin in Obstructive Sleep Apnea

M. Kosacka, I. Porębska, and A. Brzecka

Abstract

Sclerostin, a glycoprotein involved in vascular calcification, could play a role in cardiovascular disorders. Obstructive sleep apnea (OSA) is frequently associated with cardiovascular comorbidities. Thus, in this study we set out to assess the level of sclerostin in patients with OSA. Sclerostin was evaluated in the serum by ELISA method in 106 patients (43 women) with OSA of the mean age of 55 ± 10 years, BMI of 33.1 ± 7.9 kg/m^2, and apnea/hypopnea index (AHI) of 29.7 ± 18.9. There were 76 (72 %) patients with cardiovascular comorbidities in the OSA group. The results were compared with those in 49 healthy control subjects. We found that the level of sclerostin was higher in the female OSA patients than that in female controls (80.1 ± 36.5 pg/ml *vs.* 61.4 ± 24.1 pg/ml; $p < 0.05$) and it correlated with AHI ($r_s = 0.32$, $p < 0.01$) and desaturation index ($r_s = 0.34$, $p < 0.01$). Further, in OSA women with cardiovascular comorbidities, sclerostin was higher than in women without such comorbidities (87.0 ± 37.4 pg/ml *vs.* 57.3 ± 22.1 pg/ml; $p < 0.05$). In men, there were no differences in the serum sclerostin level between the OSA and control subjects, nor was there any relationship with cardiovascular diseases. In conclusion, increased serum sclerostin coincides with the severity of OSA and its cardiovascular sequelae in female patients.

Keywords

Apnea/hypopnea index • Cardiovascular risk • Obstructive sleep apnea • Sclerostin • Women

M. Kosacka (✉), I. Porębska, and A. Brzecka
Department of Pulmonology and Lung Cancer, Wroclaw Medical University, 105 Grabiszynska St., 53-439 Wrocław, Poland
e-mail: mokka113@hotmail.com

1 Introduction

Vascular calcification, a cardiovascular risk factor, is linked with increased mortality in several conditions such as diabetes or dyslipidemia

(Eyrard et al. 2014). An *in vitro* study by Zhu et al. (2011) suggested that the glycoprotein sclerostin could be involved in the vascular calcification process, which has been later confirmed *in vivo* (Balci et al. 2013; Hampson et al. 2013). Sclerostin is produced by osteocytes and has a negative influence on bone formation. It inhibits the proliferation and differentiation of osteoblastic cells (Moester et al. 2010) and induces mature osteoblast apoptosis (Sutherland et al. 2004). Sclerostin also inhibits the Wnt/β-catenin signaling pathway that is involved in embryogenesis, oncogenesis, and in the pathogenesis of osteoporosis and metabolic diseases (Mac Donald et al. 2009). The role of sclerostin has been confirmed in many bone disorders (Moester et al. 2010). Most recent studies indicate that sclerostin may also be involved in the stimulation of adipogenesis (Urano et al. 2012; Sheng et al. 2012) and may enhance the cardiovascular risk (Catalano et al. 2014; Garcia-Martin et al. 2012).

Obstructive sleep apnea (OSA) is associated with a higher incidence of cardiovascular diseases. Cardiovascular complications of OSA have a number of mechanisms such as chronic sympathetic activation, systemic inflammation, intermittent hypoxia, and sleep fragmentation (Kendzerska et al. 2014; Drager et al. 2013; Lavie and Lavie 2009). Given the link between OSA and cardiovascular complications, on the one hand, and between sclerostin and vascular calcification, on the other hand, in the present study we seek to determine the possible relationship between OSA syndrome and the level of sclerostin. We addressed the issue by assessing the serum sclerostin in OSA patients and comparing it with that in healthy control subjects.

2 Methods

This study was approved by the Bioethics Committee of Wroclaw Medical University in Wroclaw, Poland.

2.1 Patients

A total of 106 patients (F/M-43/63) of the mean age of 55 ± 10 years with newly-diagnosed OSA syndrome, as based on apnea/hypopnea index (AHI; hourly average number of apnea/hypopnea episodes longer than 10 s per hour of sleep) >5, were enrolled into the study. The mean AHI was 29.7 ± 18.9. The majority of the subjects were overweight or obese; the mean body mass index (BMI) was 33.1 ± 7.9 kg/m^2. Seventy six (72 %) patients with OSA had cardiovascular complications; one or more of the following diseases: hypertension, ischemic heart disease, diabetes, or stroke. All patients received standard treatment for cardiovascular comorbidities. A control group consisted of 49 healthy subjects, including 27 women. Subjects with a history of osteoporosis or any other bone disease were excluded from the study. The demographic data of both groups are presented in Table 1. These data stratified by gender are provided in Table 2.

All patients and control subjects underwent a nocturnal, 8-h long polysomnography with a Grass Aura Lite PSG (Warwick, RI). The following variables were continuously measured: oronasal airflow, respiratory chest and abdominal movements, and arterial oxygen saturation (SaO_2). The mean AHI, oxygen desaturation index – ODI (hourly average number of desaturation episodes), mean and minimum SaO_2 at the end of sleep apnea/hypopnea episodes were calculated.

2.2 Assessment of Sclerostin

Taking into account the known gender difference in the level of sclerostin (Catalano et al. 2014; Garcia-Martin et al. 2012) we decided to assess the peptide separately in men and women.

Samples of venous blood were taken in the morning after overnight fasting. After centrifugation with 1,467 relative centrifugal force (RCF) for 10 min, the serum was extracted and stored at −80 °C until examination. Sclerostin was

Table 1 Demographics and basic clinical data of OSA and control subjects

	OSA patients (n = 106)	Control group (n = 49)	p-value
Age (years)	55 ± 10	53 ± 10	NS
BMI (kg/m^2)	33.1 ± 7.9	30.3 ± 5.4	NS
AHI	29.7 ± 18.9	2.4 ± 1.9	<0.001
ODI	25.9 ± 20.7	2.9 ± 2.7	<0.001
CRP (mg/l)	5.5 ± 5.5	3.8 ± 3.4	NS
Sclerostin (pg/ml)	76.9 ± 32.7	66.8 ± 28.8	NS

Data are means ± SD. *AHI* apnea/hypopnea index, *ODI* oxygen desaturation index, *BMI* body mass index, *CRP* C-reactive protein

Table 2 Stratification by gender of demographic and basic clinical data, and of sclerostin level in OSA and control subjects

	Female OSA (n = 43)	Female control (n = 27)	p-value
Age (years)	58 ± 9	56 ± 10	NS
BMI (kg/m^2)	33.4 ± 8.8	32.2 ± 5.6	NS
AHI	28.5 ± 20.1	2.2 ± 1.9	<0.001
ODI	24.6 ± 20.8	2.9 ± 2.3	<0.001
CRP (mg/l)	5.4 ± 5.1	3.6 ± 2.4	NS
Sclerostin (pg/ml)	80.1 ± 36.5	61.4 ± 24.1	<0.05
	Male OSA (n = 22)	Male control (n = 63)	p-value
Age (years)	54 ± 10	49 ± 8	NS
BMI (kg/m^2)	32.9 ± 7.3	27.9 ± 4.2	<0.001
AHI	30.5 ± 18.2	2.5 ± 2.0	<0.001
ODI	26.8 ± 20.8	3.3 ± 2.9	<0.001
CRP (mg/l)	5.7 ± 5.6	4.3 ± 4.1	NS
Sclerostin (pg/ml)	74.8 ± 30.1	73.4 ± 33.1	NS

Data are means ± SD. *AHI* apnea/hypopnea, *ODI* oxygen desaturation index, *BMI* body mass index, *CRP* C-reactive protein

measured using an enzyme-linked immunosorbent assay (ELISA) method (Human SOST; R&D Systems, Minneapolis, MN), according to the manufacturer's specifications. The ELISA microplate reader from MRXe Dynex Technologies (Chantilly, VA) was used.

Additionally, C-reactive protein (CRP) and calcium were measured in the serum. Only were the subjects with a normal calcium level included into the study.

2.3 Statistical Elaboration

Data were presented as means ± SD. The intergroup differences were assessed with the Mann–Whitney U test and Spearman's *r* correlation coefficient was used to assess the relationship between variables. Statistical significance was assumed at $p < 0.05$. Statistical analysis was performed using a CSS Statistica software for Windows (ver. 5.0).

3 Results

Overall, both genders combined, sclerostin serum level tended to be higher in OSA patients than in healthy subjects (76.9 ± 32.7 *vs.* 66.8 ± 28.9; respectively, p = 0.071) and in OSA patients with cardiovascular comorbidities than without them (81.2 ± 34.4 *vs.* 66.0 ± 25.6, respectively, $p = 0.06$), with a borderline statistical significance. The differences in the level of sclerostin between OSA and control subjects, and OSA patients with and without cardiovascular comorbidities came clearly to light when the subjects were stratified by gender.

In OSA women, serum sclerostin levels were significantly higher compared with control women (Table 2). In addition, sclerostin correlated positively with BMI ($r_s = 0.30$; $p = 0.011$), AHI ($r_s = 0.32$; $p = 0.008$), and with ODI ($r_s = 0.34$; $p = 0.004$). A trend for a negative correlation of sclerostin with the mean SaO_2 also was observed ($p = 0.081$). In men, there were neither significant differences in the sclerostin level nor any sclerostin correlations with the variables above mentioned between the OSA and control subjects (Table 3).

In the female OSA patients with cardiovascular comorbidities, such as hypertension, diabetes, stroke, or ischemic heart disease ($n = 33$), sclerostin level was significantly higher than in those without them (Fig. 1). Interestingly, sclerostin concentration tended to be the highest in women specifically with ischemic heart disease (Table 4).

4 Discussion

The major finding of the study is that the serum level of sclerostin was enhanced in women suffering from OSA syndrome as compared with healthy controls. Two other important findings are that higher sclerostin levels in women are associated with a more severe course of OSA, as expressed by frequent obstructive sleep apnea/hypopnea and desaturation episodes, and with concurrent cardiovascular comorbidities. Interestingly, there seems to be a gender difference since the role of sclerostin outlined above did not concern the male OSA patients, where its changes where rather null. The corollary is that sclerostin could become a potential marker of cardiovascular sequelae in female OSA patients.

There are reports in non-OSA patients suggesting that increased sclerostin level is associated with increased cardiovascular risk. Hampson et al. (2013) have demonstrated a positive correlation between sclerostin and abdominal aorta calcification and arterial stiffness. That is in line with the findings that increased sclerostin goes in tandem with higher low-density lipoprotein cholesterol, uric acid, and homocysteine (Urano et al. 2012). Previous studies have also shown that sclerostin concentration is higher in long-lasting diabetes, diabetic macroangiopathy, and is associated with a higher level of glycated hemoglobin (Catalano et al. 2014; Garcia-Martin et al. 2012). In the present study, however, we failed to confirm the increase in sclerostin in diabetic OSA patients, which could have to do with a small number of such patients ($n = 26$).

Observations in dialysis patients also suggest that sclerostin may be involved in the vascular calcification process. However, it is debatable whether sclerostin should be considered an inhibitor or rather a promoter of this process. Balci et al. (2013) have shown that hemodialysis patients with calcified fistula have higher serum

Table 3 Correlation between sclerostin serum level and selected variables in female and male OSA patients

	Sclerostin in females		Sclerostin in males	
	r_s	p	r_s	p
Age	0.22	NS	0.18	NS
BMI	0.30	<0.05	0.01	NS
AHI	0.32	<0.01	−0.08	NS
ODI	0.34	<0.001	−0.04	NS
Mean SaO_2	−0.21	NS	0.05	NS
Min SaO_2	−0.18	NS	0.21	NS
CRP	0.18	NS	−0.19	NS

AHI apnea/hypopnea index, *ODI* oxygen desaturation index, *BMI* body mass index, *CRP* C-reactive protein

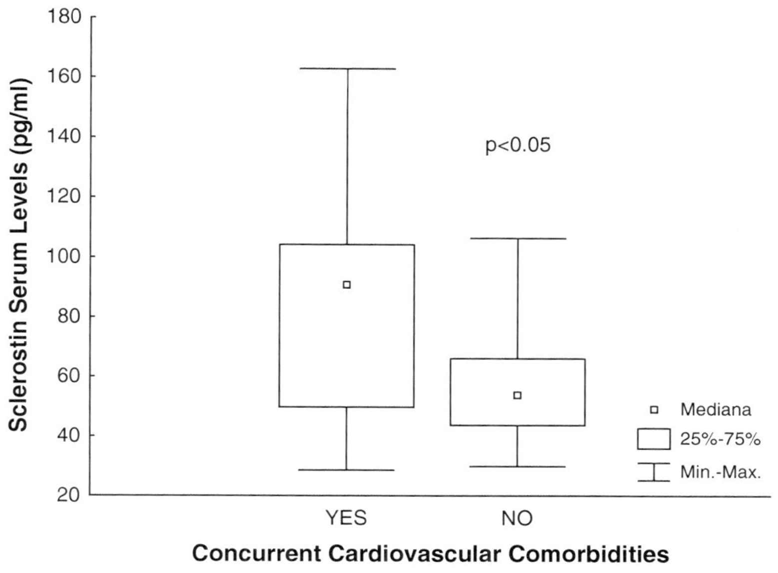

Fig. 1 Sclerostin levels in female OSA patients with and without concurrent cardiovascular comorbidity

Table 4 Sclerostin serum levels (pg/ml) in women and men with OSA syndrome without and with cardiovascular disorders

	OSA patients		
n = All/F/M	All	Females	Males
Cardiovascular comorbidity – NO (n = 30/10/20)	66.0 ± 25.6	57.3 ± 22.1	66.0 ± 19.1
Cardiovascular comorbidity -YES (n = 76/33/43)	81.2 ± 34.4	87.0 ± 37.4*	78.5 ± 33.2
Hypertension (n = 70/32/38)	79.7 ± 33.4	86.0 ± 37.7	75.2 ± 30.7
Diabetes (n = 26/11/15)	81.1 ± 34.3	83.5 ± 32.3	79.4 ± 36.6
Ischemic heart disease (n = 28/12/16)	86.9 ± 38.2	99.0 ± 37.8	77.8 ± 37.1
Stroke (n = 6/5/1)	93.4 ± 38.3	91.9 ± 42.6	119.4

Data are means ± SD; *$p < 0.05$ for the difference between the presence and absence of cardiovascular comorbidities

sclerostin levels than patients without it. In addition, these authors reported poorer arteriovenous fistula endurance in patients with enhanced sclerostin levels. Other studies, however, reported conflicting results in similar groups of patients. Claes et al. (2013) have observed that higher sclerostin levels are associated with aortic calcification in patients with chronic kidney disease. Nonetheless, in multivariate analysis, the authors revealed that lower, not higher, sclerostin levels, may be, among other factors, an independent risk factor for calcification, which actually suggests sclerostin protect against vascular calcification. In yet another study in dialysis patients, it has been found that higher sclerostin levels are associated with a lower cardiovascular mortality (Drechsler et al. 2014).

The results of the present study show that serum sclerostin levels were higher in female OSA patients than in female healthy subjects and correlated positively with the two most important OSA variables, namely AHI and ODI. These results led us to a question why this association was detected only in women. The gender role in sclerostin metabolism and sclerostin relationship with obesity and cardiovascular risk are unclear and the results of studies are contentious. Catalano et al. (2014) have shown higher sclerostin levels in type 1 diabetes in women. In contrast, Garcia-Martin et al. (2012) and Klangjareonchai et al. (2014) have shown higher sclerostin levels in type 2 diabetic or pre-diabetic males. In addition, higher sclerostin levels are also reported in postmenopausal than premenopausal women (Mirza et al. 2010), which raises the possibility of hormonal influence on sclerostin. Follicle-stimulating hormone (FSH) and estradiol have been suggested as candidate hormonal determinants of serum sclerostin in both pre- and postmenopausal women. In addition, parathyroid hormone also influence the level of sclerostin in premenopausal women (Ardawi et al. 2011). A link has been reported between sclerostin and cardiovascular risk in postmenopausal women (Hampson et al. 2013; Urano et al. 2012). In the present study, the majority of women investigated were of postmenopausal age (only 4 premenopausal women in OSA and 3 in control groups), which could explain the predominance of sclerostin effect in women.

There are gender differences in the course of OSA. Although the incidence and severity of OSA is lower in women, cardiovascular consequences may be more serious as women are at a higher risk of hypertension, endothelial dysfunction, and increased mortality (Won and Guilleminault 2015). In addition, experimental animal studies have suggested gender differences in the mechanisms of vasculature injury due to chronic intermittent hypoxia (Li et al. 2014). The present study demonstrates a positive correlation between sclerostin and BMI, but again only in women. Likewise, there are conflicting reports concerning the association between sclerostin and obesity. In line with the present findings, Urano et al. (2012) have shown that sclerostin level positively correlates with the percentage of abdominal and gynoid fat and Sheng et al. (2012) have shown a positive association of sclerostin with body weight and fat mass. No such association has been substantiated in diabetic and prediabetic subjects (Klangjareonchai et al. 2014; Garcia-Martin et al. 2012). Thus, it seems glucose metabolism may confound the interaction between sclerostin and adiposity. In addition, sexual dimorphism related to body composition could influence sclerostin association with BMI in women, but not in men. Women have a lower lean and bone mass, but a higher fat mass than men after correction for body size (Wells 2007).

It is well established that OSA syndrome is associated with increased cardiovascular risk (Kendzerska et al. 2014). The present study confirmed this relationships. The majority of our OSA patients (72 %) had cardiovascular comorbidities. We also observed that sclerostin levels were higher in female OSA patients with concurrent cardiovascular disorders than in those without it. Our results are partly in accord with the observations made by Morales-Santana et al. (2013) in diabetic non-OSA patients. Those authors have shown higher sclerostin levels in diabetic patients with atherosclerotic accompaniment than in those without it, but the results were not stratified by patient gender.

In conclusion, we believe we have shown that the serum level of sclerostin is enhanced in women suffering from OSA syndrome, and it is associated with disease severity as well as with the accompanying cardiovascular comorbidities, although the exact mechanisms of the association of sclerostin with OSA and vascular calcification remain to be determined. Nonetheless, sclerostin has a potential to become a marker of OSA severity in this category of patients.

Competing Interests The authors declare that they have no competing interests in relation to this article.

References

Ardawi MS, Al-Kadi HA, Rouzi AA, Qari MH (2011) Determinants of serum sclerostin in health pre- and postmenopausal women. J Bone Miner Res 26 (12):2812–2822

Balci M, Kirkpantur A, Turkvatan A, Mandiroglu S, Ozturk E, Afsar B (2013) Sclerostin as a new key player in arteriovenous fistula calcification. Herz 40 (2):289–297

Catalano A, Pintaudi B, Morabito N, Di Vieste G, Giunta L, Bruno ML, Cucinotta D, Lasco A, Di Benedetto A (2014) Gender differences in sclerostin and clinical characteristics in type 1 diabetes mellitus. Eur J Endocrinol 171(3):293–300

Claes KJ, Viaene L, Heve S, Meijers B, d'Haese P, Evenepoel P (2013) Sclerostin: another vascular calcification inhibitor? J Clin Endocrinol Metab 98 (8):3221–3228

Drager LF, Togeiro SM, Polotsky VY, Lorenzi-Filho G (2013) Obstructive sleep apnea: a cardiometabolic risk in obesity and the metabolic syndrome. J Am Coll Cardiol 62(7):569–576

Drechsler C, Evenepoel P, Vervloet MG, Wanner C, Ketteler M, Marx N, Floege J, Dekker FW, Brandenburg VM (2014) High levels of circulating sclerostin are associated with better cardiovascular survival in incident dialysis patients: results from the NECOSAD study. Nephrol Dial Transplant 30 (2):288–293

Eyrard S, Delanave P, Kamel S, Cristol JP, Cavalier E, SFBC/SN joined working group on vascular calcifications (2014) Vascular calcification: from pathophysiology to biomarkers. Clin Chim Acta 438:401–414

Garcia-Martin A, Rozas-Moreno P, Reyes-Garcia R, Morales-Santana S, Garcia-Fontana B, Garcia-Salcedo JA, Munoz-Torres M (2012) Circulating levels of sclerostin are increased in patients with type 2 diabetes mellitus. J Clin Endocrinol Metab 97 (1):234–241

Hampson G, Edwards S, Conroy S, Blake GM, Fogelman I, Frost ML (2013) The relationship between inhibitors of the Wnt signaling pathway (Dickkopf-1(DKK1) and sclerostin), bone mineral density, vascular calcification and arterial stiffness in post-menopausal women. Bone 56(1):42–47

Kendzerska T, Gershon AS, Hawker G, Leung RS, Tomlinson G (2014) Obstructive sleep apnea and risk of cardiovascular events and all-cause mortality: a decade –long historical cohort study. PLoS Med 11 (2):e1001599

Klangjareonchai T, Nimitphong H, Saetung S, Bhirommuang N, Samittarucksa R, Chanprasertyothin S, Sudatip R, Ongphiphadhanakul B (2014) Circulating sclerostin and irisin are related and interact with gender to influence adiposity in adults with prediabetes. Int J Endocrinol 2014:261545. doi:10.1155/2014/261545

Lavie L, Lavie P (2009) Molecular mechanisms of cardiovascular disease in OSAHS: the oxidative stress link. Eur Respir J 33:1467–1484

Li QY, Feng Y, Lin YN, Li M, Guo Q, Gu SY, Liu JL, Zhang RF, Wan HY (2014) Gender difference in protein expression of vascular wall in mice exposed to chronic intermittent hypoxia: a preliminary study. Genet Mol Res 13(4):8489–8501

Mac Donald BT, Tamai K, He X (2009) Wnt/beta-catenin signaling: components, mechanisms, and diseases. Dev Cell 17(1):9–26

Mirza FS, Padhi ID, Raisz LG, Lorenzo JA (2010) Serum sclerostin levels negatively correlate with parathyroid hormone levels and free estrogen index in postmenopausal women. J Clin Endocrinol Metab 95 (4):1991–1997

Moester MJ, Papapoulos SE, Lowik CW, van Bezooijen RL (2010) Sclerostin: current knowledge and future perspectives. Calcif Tissue Int 87(2):99–107

Morales-Santana S, Garcia-Fontana B, Garcia-Martin A, Rozas-Moreno P, Garcia-Salcedo JA, Reyes-Garcia R, Munoz-Torres M (2013) Atherosclerotic disease in type 2 diabetes is associated with an increase in sclerostin levels. Diabetes Care 36(6):1667–1674

Sheng Z, Tong D, Ou Y, Zhang H, Zhang Z, Li S, Zhou J, Zhang J, Liao E (2012) Serum sclerostin levels were positively correlated with fat mass and bone mineral density in central south Chinese postmenopausal women. Clin Endocrinol (Oxf) 76(6):797–801

Sutherland MK, Geoghegan JC, Yu C, Turcott E, Skonier JE, Winkler DG, Latham JA (2004) Sclerostin promotes the apoptosis of human osteoblastic cells: a novel regulation of bone formation. Bone 35 (4):828–835

Urano T, Shiraki M, Ouchi Y, Inoue S (2012) Association of circulating sclerostin levels with fat mass and metabolic disease—related markers in Japanese postmenopausal women. J Clin Endocrinol Metab 97(8): E1473–E1477

Wells JC (2007) Sexual dimorphism of body composition. Best Pract Res Clin Endocrinol Metab 21(3):415–430

Won C, Guilleminault C (2015) Gender differences in sleep disordered breathing: implications for therapy. Expert Rev Respir Med 9(2):221–231

Zhu D, Mackenzie NC, Millan JL, Farguharson C, MacRae VE (2011) The appearance and modulation of osteocyte marker expression during calcification of vascular smooth muscle cells. PLoS One 6(5):e19595

Advs Exp. Medicine, Biology - Neuroscience and Respiration (2016) 21: 23–30
DOI 10.1007/5584_2015_205

Published online: 29 January 2016

Atherosclerotic Vessel Changes in Sarcoidosis

I. Tuleta, S. Pingel, L. Biener, C. Pizarro, C. Hammerstingl, C. Öztürk, N. Schahab, C. Grohé, G. Nickenig, C. Schaefer, and D. Skowasch

Abstract

Sarcoidosis is a systemic granulomatous disease. Atherosclerosis is a chronic inflammatory vessel disease. The aim of our present study was to investigate whether sarcoidosis could be associated with increased risk of atherosclerotic vessel changes. Angiological analysis and blood tests were performed in 71 sarcoidosis patients and 12 matched controls in this prospective cross-sectional study. Specifically, angiological measurements comprised ankle brachial index (ABI), central pulse wave velocity (cPWV), pulse wave index (PWI), and duplex sonography of central and peripheral arteries. Sarcoidosis activity markers (angiotensin converting enzyme, soluble interleukin-2 receptor) and cardiovascular risk parameters such as cholesterol, lipoprotein(a), C-reactive protein, interleukin 6, fibrinogen, d-dimer, and blood count were analyzed in blood. We found no relevant differences in ABI, cPWV, and plaque burden between the sarcoidosis and control groups (1.10 ± 0.02 *vs.* 1.10 ± 0.02, 6.7 ± 0.5 *vs.* 6.1 ± 1.2, 53.7 % *vs.* 54.5 %, respectively). However, PWI was significantly higher in sarcoidosis patients (146.2 ± 6.8) compared with controls (104.9 ± 8.8), irrespectively of the activity of sarcoidosis and immunosuppressive medication. Except for increased lipoprotein(a) and d-dimer in sarcoidosis, the remaining cardiovascular markers were similar in both groups. We conclude that sarcoidosis is associated with increased pulse wave index, which may indicate an early stage of atherosclerosis.

I. Tuleta (✉), S. Pingel, L. Biener, C. Pizarro, C. Hammerstingl, C. Öztürk, N. Schahab, G. Nickenig, C. Schaefer, and D. Skowasch
Department of Internal Medicine II – Cardiology, Pulmonology and Angiology, University of Bonn, 25 Sigmund-Freud-St., D-53105 Bonn, Germany
e-mail: Izabela.Tuleta@ukb.uni-bonn.de

C. Grohé
Evangelische Lungenklinik Berlin-Buch, Berlin, Germany

Keywords
Arterial stiffness • Cardiovascular risk factors • Systemic inflammation • Vascular ultrasound • Vasculopathy

1 Introduction

Sarcoidosis is an inflammatory granulomatous disease which may potentially affect each organ, with a predilection for pulmonary tissue (Newman et al. 1997). Interestingly, the clinical manifestation of sarcoidosis in vascular bed is not typical for this disorder (Yildiz 2012). Sarcoidosis may be associated with pulmonary hypertension; herein primary vascular changes seem to be independent of fibrotic lung remodeling (Han et al. 2007). Little is known about the potential influence of sarcoidosis on the systemic arteries. Some reports point to increased arterial stiffness in sarcoidosis (Ardic et al. 2012; Yildiz 2012), mirroring a very early stage of atherosclerosis. However, the link between sarcoidosis and vasculopathy may depend on the activity of sarcoidosis, its duration, organ manifestation, and the drug therapy (Siasos et al. 2015; Ardic et al. 2012; Siasos et al. 2011; Han et al. 2007). The proposed mechanisms underlying atherosclerotic vessel changes in sarcoidosis involve chronic inflammation (Siasos et al. 2011; Han et al. 2007), increased oxidative stress, and dyslipidemia with HDL-cholesterol reduction (Ivanišević et al. 2012; Salazar et al. 2000), all of which is proatherogenic. Moreover, infection with *Chlamydophila pneumoniae* has been postulated as a possible trigger mechanism in the development of sarcoidosis (Choroszy-Król et al. 2014). Analogically, infection with *Chlamydia pneumoniae* may contribute to the progression of atherosclerosis (Tuleta et al. 2014).

Against the background outlined above, the purpose of the present study was to examine prospectively in a cross-sectional manner a heterogeneous population of patients with sarcoidosis in regard to the potential functional and structural artery alternations and the presence of proatherogenic factors.

2 Methods

2.1 Patient Characteristics

The research was conducted according to the principles of the Declaration of Helsinki and was approved by a local Ethics Committee, and written informed consent was obtained from all subjects. Seventy one sarcoidosis patients and twelve matched controls were enrolled into this prospective, cross-sectional study between October 2014 and January 2015. There were no relevant baseline differences between sarcoidosis and control groups in regard to age, sex, body mass index, or the presence of classical cardiovascular risk factors, except for the use of cortisone in sarcoidosis (Table 1).

Baseline characteristics of sarcoidosis patients are demonstrated in Table 2. More than one third of sarcoidosis patients demonstrated increased serum angiotensin-converting enzyme (ACE) or soluble interleukin-2 receptor (sIL-2R) concentrations as indicators of active sarcoidosis. Almost half of sarcoidosis patients was under cortisone therapy. Pulmonary sarcoidosis in stage II with reduction in lung diffusing capacity was predominant (Table 3).

2.2 Angiological Examinations

Angiological examinations included the analysis of central pulse wave velocity (cPWV), pulse wave index (PWI), ankle brachial index (ABI), and color-coded duplex sonography for the assessment of atherosclerotic plaque burden. The measurements of cPWV and PWI were conducted by means of AngE Pro8®

Table 1 Demographics and distribution of atherosclerotic risk factors in sarcoidosis patients and controls

	Age (years)	Male/ Female	BMI (kg/m^2)	Arterial hypertension	Diabetes mellitus	Never smokers	Former smokers	Current smokers	Cortisone therapy
Sarcoidosis	54.3 ± 1.5	32/71	25.0 ± 0.4	24/71	11/71	50/71	17/71	4/71	33/71
n = 71		(45.1 %)		(33.8 %)	(15.5 %)	(70.4 %)	(23.9 %)	(5.6 %)	(46.4 %)
Controls	54.8 ± 2.0	4/12	23.0 ± 1.0	5/11	0/12	6/12	3/12	3/12	0/12
n = 12		(33.3 %)		(45.5 %)	(0 %)	(50.0 %)	(25.0 %)	(25.0 %)	(0 %)*

Data are means ± Standard error of the mean (SEM) or n (%). *BMI* body mass index, $^{*}p < 0.05$

Table 2 Sarcoidosis patients' baseline characteristics

		Cortisone therapy				Stage			
	Increased ACE and/or sIL-2R	No	Low	High	Predominantly pulmonary sarcoidosis	I	II	III	IV
Sarcoidosis	25/71	38/71	29/71	4/71	60/70	8/59	36/59	12/59	3/59
n = 71	(35.2 %)	(53.5 %)	(40.8 %)	(5.6 %)	(85.7 %)	(13.6 %)	(61.0 %)	(20.3 %)	(5.1 %)

Data are presented as n (%)

Table 3 Reduction of lung diffusing capacity in predominantly pulmonary sarcoidosis

	No	Low	Moderate	Severe
Sarcoidosis	30/71	20/71	14/71	7/71
n = 71	(42.3 %)	(28.2 %)	(19.7 %)	(9.9 %)

Data are presented as n (%). *ACE* angiotensin converting enzyme, *sIL-2R* interleukin-2 receptor, cortisone therapy: low = ≤15 mg prednisolone/day, high= > 15 mg prednisolone/day. Reduction of lung diffusing capacity expressed as carbon monoxide transfer factor (TLCO): no = TLCO ≥ 80 %, low = 65 % ≤ TLCO < 80 %, moderate = 50 % ≤ TLCO < 65 %, severe = TLCO < 50 %

(Sonotechnik Austria, Maria Rain, Austria). The cPWV greater than 12 m/s was considered pathological (Schaefer et al. 2015). The PWI was calculated as a mean ratio of the maximal pulse wave amplitude of upper and lower extremity multiplied by the summit time, as previously reported (Pizarro et al. 2015; Schaefer et al. 2015). It is a very sensitive and blood pressure independent parameter for the diagnosis of early vessel changes. The ABI was calculated as a mean ratio of the systolic blood pressure of the tibial posterior and dorsal pedis arteries and the systolic blood pressure of both arms. An ABI < 0.9 and ABI > 1.3 indicated peripheral artery disease and increased vascular stiffness in the form of medial sclerosis, respectively. Atherosclerotic arterial disease (AAD) was defined as at least one atherosclerotic plaque in central or peripheral arteries in the duplex sonography.

2.3 Blood Tests

Blood samples were collected at the time of angiological measurements. The following sarcoidosis-related parameters were assessed: ACE, sIL-2R and different cardiovascular risk factors such as total cholesterol, HDL-cholesterol, LDL-cholesterol, lipoprotein(a), C-reactive protein (CRP), interleukin 6, fibrinogen, d-dimer, and blood count.

2.4 Statistical Analysis

Differences between the mean values calculated for both groups were assessed by a *t*-test. Categorical variables were compared by the Chi-squared test. A p-value of <0.05 was considered as statistically significant.

3 Results

The ABI, plaque load, and cPWV did not show any significant differences between the sarcoidosis and control groups. The only angiological parameter that demonstrated relevant differences between the groups was the PWI (Table 4). The measurements of potential cardiovascular risk factors in blood (Fig. 1) revealed significant increases in lipoprotein(a) and d-dimer in the sarcoidosis group. We found also a tendency toward higher CRP blood concentrations in the sarcoidosis patients, albeit the difference failed to reach statistical significance compared with the control subjects. A stratification by different sarcoidosis subgroups, such as active sarcoidosis with either increased ACE/sIL-2R or cortisone therapy, predominant pulmonary or extrapulmonary sarcoidosis, and progressive sarcoidosis stages III and IV failed to substantiate appreciable differences in angiological and blood results between the sarcoidosis patients and control subjects (data not shown).

4 Discussion

Our present angiological results did not show any increased structural vessel changes in the form of plaque development with or without consecutive flow limitation in sarcoidosis patients compared with controls. Likewise, established functional tests, such as the cPWV did not reveal any significant differences in sarcoidosis *vs.* control subjects. The only measurement that demonstrated

Table 4 Angiological analysis

	cPWV	PWI	ABI	AAD
Sarcoidosis	6.7 ± 0.5	146.2 ± 6.8*	1.1 ± 0.02	36/67
n = 71				(53.7 %)
Control	6.1 ± 1.2	104.9 ± 8.8	1.1 ± 0.02	6/11
n = 12				(54.5 %)

Data are presented as means ± Standard error of the mean (SEM) or n (%). *cPWV* central pulse wave velocity (m/s), *PWI* pulse wave index, *ABI* ankle brachial index, *AAD* atherosclerotic arterial disease, *p < 0.05 for PWI between both groups

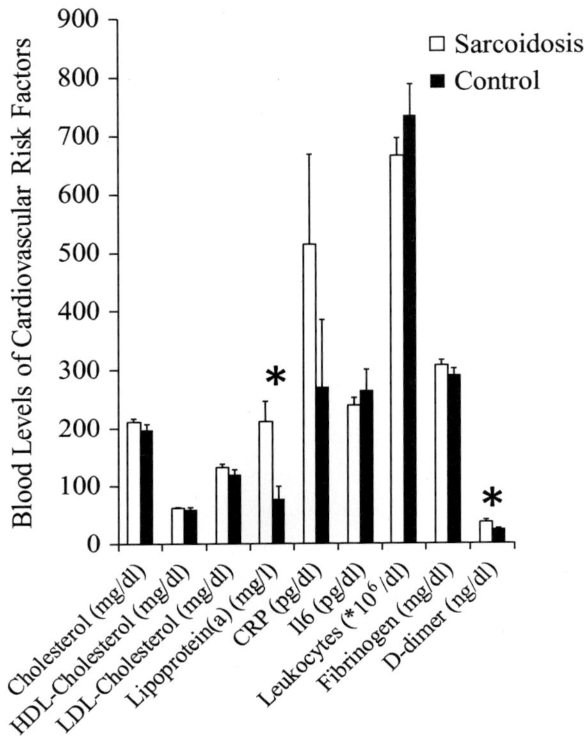

Fig. 1 Potential cardiovascular risk factors in sarcoidosis and control groups. Except for lipoprotein(a) and d-dimer there were no relevant differences in other blood factors; *p < 0.05

significant changes was elevated PWI in sarcoidosis patients. Assuming that PWI is a very sensitive marker of functional vessel changes leading to the manifestation of atherosclerosis down the road, our study, in line with some other reports (Ardic et al. 2012; Yildiz 2012), suggests that sarcoidosis patients may suffer from pathological vessel changes. Moreover, analysis of cardiovascular risk factors in blood showed significantly elevated levels of lipoprotein(a) and d-dimer, and a trend toward higher CRP concentrations in sarcoidosis patients in comparison with controls. Since lipoprotein(a) is an independent risk factor for the development of cardiovascular diseases *via* its prothrombotic/antifibrinolytic and proinflammatory effects (Nordestgaard et al. 2010; Sotiriou et al. 2006), the above result could at least partly explain the arterial alternations in sarcoidosis patients. Increased levels of lipoprotein(a) have also been reported in patients with other than sarcoidal granulomatous diseases such as rheumatoid arthritis. Such patients could be at a high risk of developing of a cardiovascular disease (Govindan et al. 2015). Likewise, higher concentrations of d-dimer markers in blood, found in our present work and in other studies (Shorr and Hnatiuk 2000) are associated with increased arterial stiffness and faster pulse wave propagation (Wykretowicz et al. 2012). A tendency toward increased CRP values in sarcoidosis noted in the present study may suggest that augmented inflammatory response underlies functional vessel changes. Indeed, pronounced inflammation has been detected in sarcoidosis patients characterized by a decreased arterial elasticity (Siasos et al. 2011; Han et al. 2007).

In contrast to other studies, activity of sarcoidosis defined either as increases in ACE/sIL-2R in blood or cortisone therapy, stages of pulmonary sarcoidosis, and as pulmonary or extrapulmonary manifestations failed to influence the findings of the present study. This discrepancy may be explained by a relatively small, statistically considering, number of patients with severe sarcoidosis characterized by increased sarcoidosis blood markers, advanced pulmonary stages, or therapy with high prednisolone doses in our patients.

In conclusion, this study suggests pathological arterial changes in sarcoidosis patients associated with increased cardiovascular risk factors such as lipoprotein(a) and d-dimer in blood. However, other variables describing functional capacity of arteries remain to be further explored to settle the issue of atherosclerotic arterial alterations in sarcoidosis.

Acknowledgments We gratefully acknowledge Sonotechnik Austria for providing us with the AngE Pro8®. We would like to thank all patients who participated in the study and the Sarkoidose-Netzwerk for support.

Conflicts of Interest The authors declare no conflicts of interest in relation to this article.

References

Ardic I, Yarlioglues M, Dogdu O, Buyukoglan H, Kanbay A, Akpek M, Bol C, Yuksel M, Akkaya E, Vuruskan E, Kaya MG (2012) Assessment of aortic elastic properties in patients with sarcoidosis. Blood Press 21:286–292

Choroszy-Król I, Frej-Mądrzak M, Hober M, Sarowska J, Jama-Kmiecik A (2014) Infections caused by Chlamydophila pneumoniae. Adv Clin Exp Med 23:123–126

Govindan KP, Basha S, Ramesh V, Kumar CN, Swathi S (2015) A comparative study on serum lipoprotein (a) and lipid profile between rheumatoid arthritis patients and normal subjects. J Pharm Bioallied Sci 7:S22–S25

Han MK, McLaughlin VV, Criner GJ, Martinez FJ (2007) Pulmonary diseases and the heart. Circulation 116:2992–3005. Review

Ivanišević J, Kotur-Stevuljević J, Stefanović A, Jelić-Ivanović Z, Spasić S, Videnović-Ivanov J, Vučinić-Mihailović V, Ilić J (2012) Dyslipidemia and oxidative stress in sarcoidosis patients. Clin Biochem 45:677–682

Newman LS, Rose CS, Maier LA (1997) Sarcoidosis. N Engl J Med 336:1224–1234

Nordestgaard BG, Chapman MJ, Ray K (2010) Lipoprotein(a) as a cardiovascular risk factor: current status. Eur Heart J 31:2844–2853

Pizarro C, Schaefer C, Kimeu I, Pingel S, Horlbeck F, Tuleta I, Nickenig G, Skowasch D (2015) Underdiagnosis of obstructive sleep apnoea in peripheral arterial disease. Respiration, (Epub ahead of print)

Salazar A, Maña J, Fiol C, Hurtado I, Argimon JM, Pujol R, Pinto X (2000) Influence of serum amyloid A on the decrease of high density lipoprotein-cholesterol in active sarcoidosis. Atherosclerosis 152:497–502

Schaefer CA, Adam L, Weisser-Thomas J, Pingel S, Vogel G, Klarmann-Schulz U, Nickenig G, Pizarro C, Skowasch D (2015) High prevalence of peripheral arterial disease in patients with obstructive sleep apnoea. Clin Res Cardiol 104:719–726

Shorr AF, Hnatiuk OW (2000) Circulating D dimer in patients with sarcoidosis. Chest 117:1012–1016

Siasos G, Paraskevopoulos T, Gialafos E, Rapti A, Oikonomou E, Zaromitidou M, Mourouzis K, Siasou G, Gouliopoulos N, Tsalamandris S, Vlasis K, Stefanadis C, Papavassiliou AG, Tousoulis D (2015) Vascular function and ocular involvement in sarcoidosis. Microvasc Res 100:54–58

Siasos G, Tousoulis D, Gialafos E, Oikonomou E, Zaromitidou M, Aggeli C, Korompelis P, Kallianos A, Rapti A, Zisimos K, Marinos G, Stefanadis C, Papavassiliou AG (2011) Association of sarcoidosis with endothelial function, arterial wall properties, and biomarkers of inflammation. Am J Hypertens 24:647–653

Sotiriou SN, Orlova VV, Al-Fakhri N (2006) Lipoprotein (a) in atherosclerotic plaques recruits inflammatory cells through interaction with Mac-1 integrin. FASEB J 20:559–561

Tuleta I, Reek D, Braun P, Bauriedel G, Nickenig G, Skowasch D, Andrié R (2014) Influence of intimal Chlamydophila pneumoniae persistence on

cardiovascular complications after coronary intervention. Infection 43:51–57
Wykretowicz J, Guzik P, Krauze T, Marciniak R, Komarnicki M, Piskorski J, Wysocki H, Wykretowicz A (2012) Fibrinogen and d-dimer in contrasting relation with measures of wave reflection and arterial stiffness. Scand J Clin Lab Invest 72:629–634
Yildiz M (2012) Arterial distensibility in chronic inflammatory rheumatic disorders. Open Cardiovasc Med J 4:83–88

Advs Exp. Medicine, Biology - Neuroscience and Respiration (2016) 21: 31–38
DOI 10.1007/5584_2015_206

Published online: 29 January 2016

Asthma and COPD: Similarities and Differences in the Pathophysiology, Diagnosis and Therapy

Josef Yayan and Kurt Rasche

Abstract

Asthma and chronic obstructive pulmonary disease (COPD) are two of the most common chronic lung diseases worldwide. Distinguishing between these different pulmonary diseases can be difficult in practice because of symptomatic similarities. A definitive diagnosis is essential for correct treatment. This review article presents the different symptoms of these two chronic inflammatory lung diseases following a selective search of the PubMed database for relevant literature published between 1996 and 2012. While cough occurs in both diseases, asthmatics often have a dry cough mainly at night, which is often associated with allergies. In contrast, COPD is usually caused by years of smoking. Paroxysmal dyspnea, which occurs in asthma, is characterized by shortness of breath, while in COPD it occurs during physical exertion in early stages and at rest in later stages of the disease. Asthma often begins in childhood or adolescence, whereas COPD occurs mainly in smokers in later life. It is possible to live with asthma into old age, whereas the life expectancy of patients with COPD is significantly limited. Currently, there is no general curative treatment for either disorder.

Keywords

Asthma • COPD • Detection • Symptoms • Treatment

1 Introduction

Bronchial asthma and chronic obstructive pulmonary disease (COPD) are among the most common chronic diseases of the lungs. They have similar symptoms that include shortness of breath, cough, and sputum, and these similarities can often cause difficulties in making the correct

J. Yayan (✉) and K. Rasche
Department of Internal Medicine, Division of Pulmonary, Allergy and Sleep Medicine, HELIOS Clinic Wuppertal, Witten/Herdecke University, 40 Heusnerstrasse, 42283 Wuppertal, Germany
e-mail: josef.yayan@hotmail.com

diagnosis for physicians in everyday practice (Carolan and Sutherland 2013). These lung diseases differ from each other mainly in terms of pathogenesis, disease progression, prognosis, and treatment options (Chang and Mosenifar 2007). It is important to distinguish between asthma and COPD for proper treatment, but early diagnosis and initiation of treatment remain a major challenge for even the experienced clinician and some pulmonologists (Rothe 2012). The correct guideline-based diagnosis and appropriate pharmacological treatment of these two common diseases is extremely important.

Against this background, the present study investigates the various clinical symptoms of asthma and COPD to improve their rapid early detection. A search of previous relevant publications from 1996 to 2012 in the PubMed database was conducted, and their results were analyzed. Using this information, improvements can be made to the guidelines for drug treatment options for these diseases. Another major objective of this study is to discover new distinctive features in clinical symptoms, diagnosis, and treatment of both asthma and COPD.

2 Allergic Asthma

Allergic asthma is characterized as a form of chronic inflammation of the lungs and in particular by immunoglobulin-E (IgE)-mediated hypersensitivity of bronchi to aero-allergens, with increased mucus secretion. While the airway inflammation induced by allergen exposure is almost always eosinophilic, most non-allergic asthma is also associated with eosinophils. The result of this progressive chronic bronchial inflammatory process is called allergic asthma (Postma and Kerstjens 1998). The airway is obstructed by increased mucus secretion, with irreversible structural remodeling of the airways (Aoshiba and Nagai 2004). In addition, the bronchial walls can thicken over the course of the disease due to an increase in mucous hypersecretion. The structural remodeling of bronchi is largely responsible for the long-term narrowing of airways and thus it underlies the major disease symptoms, such as shortness of breath, cough, and sputum in both severe asthma and also in COPD (Górska et al. 2009). Bronchi of the lower airways are particularly hypersensitive to cold air, perfumes, cigarette smoke, and other non-specific stimuli in the air, which is referred to as airway hyperresponsiveness (Miglino et al. 2012; Bleecker 2004). Asthma can often be treated by individually tailored drug therapy to manage symptoms, but it currently is still an incurable, chronic disease (Cukic et al. 2012).

Often, when given the diagnosis of asthma, patients ask 'Why me?' The fact is that according to the World Health Organization more and more people suffer from asthma, with over 300 million people with asthma in the world and approximately 250,000 associated deaths (Cukic et al. 2012). There are certain risk factors that contribute to the severity of asthma that are considered the casual factors for the dramatic increase in the incidence of this disease. In addition, genetic predisposition and certain environmental factors probably are at play in the appearance of asthma (Kaneko et al. 2013; Bleecker 2004).

3 Chronic Obstructive Pulmonary Disease (COPD)

According to the WHO, COPD is among the most common causes of death worldwide, with approximately six million people affected in Germany and 16 million affected in the United States. These figures may represent low estimates, because there is evidence of missing or delayed diagnoses of COPD (Bleecker 2004). The main risk factor for COPD is tobacco smoke (Sutherland and Martin 2003). Most patients suffering from COPD are either active smokers, passive smokers, or former smokers (Kardos et al. 2006). Accordingly, COPD is also referred to, in common parlance, as smoker's cough. In some cases, people have a hereditary genetic predisposition to develop COPD, which is related to alpha-1-protease inhibitor deficiency (Kaneko et al. 2013). The lack of the enzyme results in unopposed proteolytic damage to the alveolar capillary membrane and does not rely on amplified inflammation of the lung caused by a mutation in the protein of alpha-1-protease inhibitor (Kardos et al. 2006; Sutherland and Martin 2003). COPD leads to recurrent infections of the lungs, which accelerates the destruction of lung tissue. In later stages of the disease, comorbidities such as hypertension, coronary artery disease, diabetes, lung cancer, or depression worsen symptoms.

4 Similarities and Differences Between Asthma and COPD

The classical symptom of COPD is exertional dyspnea. Cough and sputum are inconsistent symptoms of COPD especially in ex-smokers (Bleecker 2004). Chronic cough occurs almost universally in smokers without airflow obstruction, which underlines the definition of chronic bronchitis. It usually goes away in COPD patients who stop smoking, which constitutes a majority of patients. Bronchial asthma causes similar symptoms. In both diseases, there is chronic inflammation with cellular and structural changes, known as airway remodeling with reconstruction, and these structural changes lead to the thickening of airway walls, which facilitates airway constriction and, consequently, airflow restriction (Nakawah et al. 2013; Miglino et al. 2012; Górska et al. 2009). Nonetheless, there are key differentiators between the two diseases. The pattern of infiltrated cells and structural changes are different. In asthma, CD4, T-lymphocytes, eosinophils, and mast cells are the predominant cells involved, while in COPD, CD8, T-lymphocytes, and macrophages are predominantly involved. In severe cases of both asthma and COPD, there also can be infiltration of neutrophils into the airway walls. The thickening of smooth muscles dominates in large airways in severe asthma and in small airways in COPD (Sköld 2010; Welte and Groneberg 2006; Aoshiba and Nagai 2004). Structural similarities of the reconstruction process in asthma and COPD consist of the thickening of bronchial wall and airway mucosa, the proliferation of mucus-producing cells, and of the luminal narrowing of airways by increased secretion of inflammatory cells and mucus (Górska et al. 2009). Typical structural alterations in COPD include epithelial changes due to cell metaplasia, mucous membrane changes, and fibrosis of airway walls. The destruction and fibrosis of the alveolar wall is indicative of COPD, but not of asthma (Aoshiba and Nagai 2004). Another difference is the patient age at onset, which is often in childhood and usually before 40 years of age in asthmatics, while COPD usually occurs after 50 years of age. There is often a causal relationship to allergic diseases among asthmatics (Yawn 2009). Occasionally, no allergens are identified in patients with asthma, a situation which constitutes a non-allergic or intrinsic asthma. That could also mean that allergens remain unrecognized, rather than absent, in such patients. It is more difficult to distinguish between asthma and COPD when the patient has a long history of smoking, as the patient may suffer from both diseases (Kardos et al. 2006). Previously, post beta-agonist reversibility of bronchial obstruction was considered a distinguishing feature of asthma from COPD, but this is no longer valid based on recent clinical experience (Nakawah et al. 2013; Bleecker 2004). COPD patients can show a degree of bronchial reversibility, and conversely asthma patients can show poor post beta-agonist reversibility, having a fixed obstruction similar to that appearing in COPD. Thus, in differentiating the two diseases, the patient history and clinical evaluation ought to be taken into account (Kardos et al. 2006; Welte and Groneberg 2006; Bleecker 2004). The most common reason for the lack of bronchial reversibility in asthma is that baseline spirometry is normal. To fully examine the test post beta-agonist reversibility, a trial of corticosteroid is needed. A history of tobacco smoking exacerbates the symptoms not only in asthma and COPD, but also in a range of other lung ailments. Tobacco smoking causes atherosclerosis, resulting in afflictions such as myocardial infarction, depression, osteoporosis, and diabetes (Bleecker 2004).

5 Differences in Diagnosis Between COPD and Asthma

The task of the lung is to exchange gas. Oxygen is breathed from the ambient air *via* airways into the air sacs where it is exchanged with carbon dioxide coming from cells, which is then exhaled from the body. There are various investigative methods to measure lung volume, such as spirometry and body plethysmography (Abramson et al. 2012; Mosenifar 2009). Table 1 shows the most important differentiating factors in the major characteristics of asthma and COPD.

In both diseases, respiratory lung function is reduced. In COPD, airway resistance persistently increases, and while this can also be the case in asthma, airway resistance typically increases during an acute exacerbation of asthma. Inhaling beta2-sympathomimetics may reduce airway resistance and improve lung function in asthmatics. Such a drug effect occurs only to a small extent in patients with COPD. While spirometry gives often normal readings in an attack-free period in asthmatics, such findings are seldom in COPD patients, with a forced expiratory volume in one second (FEV1) less than 70 % of forced vital capacity (FVC) (Tandon et al. 2013; Miravitlles et al. 2012; Oga et al. 2010; Mosenifar 2009; Silvestri et al. 2008; Guerra 2005). The flow rate of exhaled air during the forced exhalation depends on airway resistance and is referred to as peak flow. It can be measured by spirometry or with a peak flow meter. The measurement is useful for asthmatics, who can be administered drugs during large peak flow fluctuations, which do not frequently occur in COPD (Mishima 2009; Pauwels 2004).

Radiographic examination shows no typical signs for asthma. In COPD, on the other side, signs of lung emphysema are present. Emphysema is inferred from vascular paucity or bullae as well as from increased radiolucency of lungs, flattening of the diaphragm, extended intercostal spaces, and rarefaction of blood vessels in the lung periphery seen in radiological examination (Chang and Mosenifar 2007; Martinez et al. 2005); the signs signifying the presence of lung hyperinflation, although hyperinflation can occur without emphysema in pulmonary diseases like asthma. Table 2 lists basic differences in the diagnosis and treatment of asthma and COPD.

In asthma patients, eosinophils are predominantly detected in sputum, whereas in COPD patients neutrophils are found (Pauwels 2004; Sutherland and Martin 2003). A noteworthy difference in drug therapy is that glucocorticoids are highly effective in asthmatics, but are of a meager effect in COPD (Bumbacea and Bogdan 2011; Morice et al. 2008; Kardos et al. 2006; Decramer and Selroos 2005; Löfdahl et al. 2005).

Table 1 Clinical differences between bronchial asthma and COPD

	Asthma	COPD
Symptoms		
Shortness of breath	During acute exacerbation	First during exercise; later at rest
Cough	Dry cough, often at night	Smoker's cough; especially in the morning with mucus
Expectoration	Crystal clear, very tenacious	Clear, yellowish
Age of onset	Any age, more common in children and adolescents, occasionally at age 70 and over	Over 40 years
Disease onset	Often suddenly	slowly, then chronic progressing bronchitis
Trigger	Infection, cold, stress, irritants, exercise, neurodermitis	Tobacco smoke
Cause	Pollen, house dust mites, animal hair, inheritance	Smoking, air pollution
Life expectation	Life possible into old age	Significantly shortened, typically very slow decline

COPD chronic obstructive pulmonary disease

Table 2 Diagnostic and treatment differences between asthma and COPD

	Asthma	COPD
Spirometry		
Respiratory function	Reduced	Reduced
Breathing resistance	Increased during acute exacerbation; reduced by beta-2 sympathomimetics	Increased; hardly reduced by beta-2 sympathomimetics
Spirometry findings	Often normal in the absence of acute exacerbation	FEV1/FVC ratio less than the lower limit of normal
Peak flow	Large fluctuations; reduced in the morning	Very small fluctuations
Radiographic examination	No typical change	Signs of pulmonary emphysema: increased lung radiolucency, diaphragm flattening, extended intercostal spaces, rarefaction of blood vessels in lung periphery
Laboratory	Eosinophils, immunoglobulin-E	Neutrophils
Sputum	Eosinophils	Neutrophils
Glucocorticoids	Great effect	Little effect

COPD chronic obstructive pulmonary disease, *FEV1* forced expiratory volume in one second, *FVC* forced vital capacity, *VC* vital capacity

6 Discussion

The chronic inflammatory respiratory diseases asthma and COPD are found worldwide, but their prevalence is underestimated (Athanazio 2012). The COPD-related mortality is on the rise. It is the fourth leading cause of death around the world (Cukic et al. 2012). Accordingly, COPD and asthma are major causes of chronic morbidity and constitute a global health challenge. It is expected that the prevalence of COPD will continue to grow in the years to come as the life expectancy globally increases; the diseases appears along the aging process (Bleecker 2004). Additionally, there is a long time lag between the initiation of smoking in early life and the late onset of disease. A misdiagnosis of COPD or asthma leads to inadequate treatment and raises health care costs (Spencer and Krieger 2013; Kuebler et al. 2008).

Differentiation between asthma and COPD in their early phases is essential for the introduction of therapeutic measures. Apart from a higher economic burden, COPD has a less favorable prognosis than asthma and is associated with greater morbidity and mortality (Decramer and Selroos 2005). However, due to a high prevalence of both diseases and their common pathophysiological processes, some patients may manifest similar symptoms making the diagnosis ambiguous or difficult (Athanazio 2012). While the most common symptoms, such as shortness of breath, cough, and sputum can singly or simultaneously occur in asthma or COPD, they differ in the way they appear. Other disorders may also be behind these cardinal symptoms, such as cardiovascular or ear, nose, and throat diseases. A simultaneous presence of infectious comorbidities, manifesting a similar clinical picture, can further obscure the diagnosis of asthma or COPD. A cardinal symptom of COPD, which may be accompanied by cough, phlegm, and wheezing, is shortness of breath during exertion. Older people often experience such shortness of breath due to deconditioning or obesity. Cough and sputum is so widespread in smokers that it is often viewed inattentively. Wheezing occurs frequently in asthmatics, but the casual factor could also be viral infections of airways (Martinez et al. 2005; Bleecker 2004). Further, asthma and COPD may coexist. While asthma alone typically occurs in younger patients, the onset of both overlapping diseases may appear at a similar later age. The hospitalization rate in patients with the overlapping diseases is almost twice as high as that for the isolated occurrence of asthma and COPD (Andersén et al. 2013).

Studies suggest that asthma and COPD are pathophysiologically and clinically different. The extent to which allergies and heredity can contribute to the development of asthma is not fully understood. COPD, however, is mainly caused by tobacco smoking, although a certain predisposition may also play a causative role (Bleecker 2004). According to some hypotheses, asthma and COPD either have a common origin with different phenotypic presentations or are caused by the interaction between endogenous and exogenous factors. There are apparent differences in the two diseases concerning the majority of inflammatory cells and mediators, albeit a number of commonalities in the inflammation process have been found. Generally speaking, COPD is diagnosed at a rather advanced age. A diagnosis of asthma from pulmonary function tests implies a completely reversible post beta-agonist narrowing of airways. Hyperinflation at rest makes a diagnosis of COPD likely. COPD should also be considered in cases of limited diffusion capacity, since these measurements are usually normal or even elevated in asthmatics. A reduced elasticity of lungs is a hallmark of COPD, and a pathophysiological enlargement of airspaces with destruction of the airway wall is especially notable with emphysema. In younger patients, atopy is indicative of asthma diagnosis (Chang and Mosenifar 2007; Martinez et al. 2005; Sciurba 2004). Other triggers, such as dust, influenza infection, climbing stairs, smoking, chemicals, or pollen, may contribute to the development of asthma and COPD, with similar triggering frequency in both diseases (Aydin et al. 2013). Preventive actions against these triggers are undertaken more frequently in patients with asthma. In contrast, vaccination rates for influenza and pneumococcal pneumonia are significantly higher in patients with COPD. It seems that the education of these two patient groups should focus on triggers and strategies on how to manage the disease. Both diseases have common features, such as obstruction, inflammation, and hypersensitivity of airways. Nonetheless, since the inherent underlying mechanisms are different, they are considered two different pathologies, which require different assessment, monitoring, and treatment (Hoshino et al. 2009).

Recently, indications have emerged that viral respiratory infections are a major cause of exacerbations of asthma and COPD. Much of the increase in morbidity, mortality, and health care costs are caused by acute exacerbations. There is no effective drug therapy for the prevention or treatment of virus-induced exacerbations. Developing new therapies requires a better understanding of how the molecular and cellular mechanisms shape the role of viral infections in exacerbations of asthma and COPD (Mallia et al. 2007). The frequency of exacerbations is similar in asthma and COPD and it is related to the severity of disease. Common causes are viral or bacterial infections, pollution, and increased allergen exposure. Eosinophilia and neutrophilia have also been associated with exacerbations and avoidance of causative factors reduces the exacerbation rate in both diseases. Pharmacological prevention of exacerbations has been demonstrated in asthma using monoclonal anti-IgE with long-acting inhaled beta2 agonists in addition to inhaled corticosteroid therapy. Exacerbations in COPD have been reduced with inhaled corticosteroids, long-acting inhaled beta2 agonists, or a combination of both with the possible addition of long-acting inhaled anticholinergics (Pauwels 2004).

The gold standard for the diagnosis of asthma and COPD is spirometry, but this is seldom used in practice. For this reason, COPD is often underdiagnosed. While the spirometry findings for differentiating between asthma and COPD are ignored in everyday practice, they are helpful in difficult or questionable cases (Abramson et al. 2012; Miravitlles et al. 2012). Although diagnostic guidelines exist for both diseases, differentiation between asthma and COPD is not full clear. Therefore, a better awareness of differences in spirometry between the two diseases would help promote the optimal treatment (Tinkelman et al. 2006). One study has attempted to develop a questionnaire for the distinction between asthma and COPD (Beeh et al. 2004). By collecting data of adult patients with a diagnosis of asthma or COPD from a pulmonary specialist practice, a simple quantitative questionnaire has been created to support a diagnosis of COPD with a high score and asthma with a low score. The questionnaire results were compared against the diagnoses made at a doctor's office according to the guidelines of the Global Initiative for Asthma (GINA) and the Global Initiative for Chronic Obstructive Lung Diseases (GOLD), including skin tests, spirometry, and reversibility. The questionnaire appears to facilitate the differentiation between COPD and asthma in daily clinical practice when applied to the former patients (Miravitlles et al. 2012; Beeh et al. 2004). Further studies are needed to confirm these initial observations.

Recent studies have unraveled additional facts for a close epidemiological and clinical relationship between asthma and COPD. Adult patients with asthma are 12 times more likely to develop COPD over the course of their lives than subjects without asthma. Consequently, early identification of patients with signs of asthma could be considered as an indication of a predisposition for the later development of COPD. These findings may have an effect on the prevention of COPD (Guerra 2005). In any case, the exact relationship between the chronic lung diseases asthma and COPD remains to be explored (Jenkins et al. 2005). An enhanced understanding of the pathophysiology of multiple obstructive pulmonary diseases is required to promote efficient, early, and optimal treatment of asthma and to prevent its progression to COPD.

7 Conclusions

Although asthma and COPD have many symptomatic similarities, the causes and pathophysiology of the diseases remain unclear. While COPD is associated with an abnormal radiographic examination, asthma has normal airway resistance between exacerbations. COPD is commonly known as smoker's cough. In asthma, eosinophils are predominantly detected in sputum. Asthma is often caused by animal hair and triggered by neurodermatitis. Differences in the inflammation process must be clarified in both diseases with more extensive research. A disturbed gas exchange that occurs in COPD, but not in asthma, also needs to be investigated, alongside a rather common occurrence of other systemic comorbidities being significantly increased in COPD patients. New diagnostic tools and blood markers help identify and differentiate the two lung diseases, which ensures early treatment. The development of new strategies for better patient-oriented treatment of respiratory diseases is sought.

Competing Interests The authors report no conflicts of interest in relation to this work.

References

Abramson MJ, Schattner RL, Sulaiman ND, Del Colle EA, Aroni R, Thien F (2012) Accuracy of asthma and COPD diagnosis in Australian general practice: a mixed methods study. Prim Care Respir J 21:167–173

Andersén H, Lampela P, Nevanlinna A, Säynäjäkangas O, Keistinen T (2013) High hospital burden in overlap syndrome of asthma and COPD. Clin Respir J 7:342–346

Aoshiba K, Nagai A (2004) Differences in airway remodeling between asthma and chronic obstructive pulmonary disease. Clin Rev Allergy Immunol 27:35–43

Athanazio R (2012) Airway disease: similarities and differences between asthma, COPD and bronchiectasis. Clinics (Sao Paulo) 67:1335–1343

Aydin Ö, Celik GE, Önen ZP, Yilmaz İ, Özdemir SK, Yildiz Ö, Mungan D, Demirel YS (2013) Triggers of asthma and COPD: are they different? Allergol Immunopathol (Madr) 41:30–36

Beeh KM, Kornmann O, Beier J, Ksoll M, Buhl R (2004) Clinical application of a simple questionnaire for the differentiation of asthma and chronic obstructive pulmonary disease. Respir Med 98:591–597

Bleecker ER (2004) Similarities and differences in asthma and COPD. The Dutch hypothesis. Chest 126:93S–95S

Bumbacea D, Bogdan MA (2011) Update in pneumology – focus on asthma and COPD. Maedica (Buchar) 6:339–340

Carolan BJ, Sutherland ER (2013) Clinical phenotypes of chronic obstructive pulmonary disease and asthma: recent advances. J Allergy Clin Immunol 131:627–634

Chang J, Mosenifar Z (2007) Differentiating COPD from asthma in clinical practice. J Intensive Care Med 22:300–309

Cukic V, Lovre V, Dragisic D, Ustamujic A (2012) Asthma and chronic obstructive pulmonary disease

(COPD) – differences and similarities. Mater Sociomed 24:100–105

Decramer M, Selroos O (2005) Asthma and COPD: differences and similarities. With special reference to the usefulness of budesonide/formoterol in a single inhaler (Symbicort) in both diseases. Int J Clin Pract 59:385–398

Górska K, Krenke R, Kosciuch J, Korczynski P, Zukowska M, Domagala-Kulawik J, Maskey-Warzechowska M, Chazan R (2009) Relationship between airway inflammation and remodeling in patients with asthma and chronic obstructive pulmonary disease. Eur J Med Res 14(Suppl 4):90–96

Guerra S (2005) Overlap of asthma and chronic obstructive pulmonary disease. Curr Opin Pulm Med 11:7–13

Hoshino T, Toda R, Aizawa H (2009) Pharmacological treatment in asthma and COPD. Allergol Int 58:341–346

Jenkins CR, Thompson PJ, Gibson PG, Wood-Baker R (2005) Distinguishing asthma and chronic obstructive pulmonary disease: why, why not and how? Med J Aust 183:S35–S37

Kaneko Y, Yatagai Y, Yamada H, Iijima H, Masuko H, Sakamoto T, Hizawa N (2013) The search for common pathways underlying asthma and COPD. Int J Chron Obstruct Pulmon Dis 8:65–78

Kardos P, Brutsche M, Buhl R, Gillissen A, Rabe KF, Russi EW, Sauer R, Worth H, Menz G (2006) Combination of asthma and COPD: more frequent as considered to be? Pneumologie 60:366–372

Kuebler KK, Buchsel PC, Balkstra CR (2008) Differentiating chronic obstructive pulmonary disease from asthma. J Am Acad Nurse Pract 20:445–454

Löfdahl CG, Ericsson A, Svensson K, Andreasson E (2005) Cost effectiveness of budesonide/formoterol in a single inhaler for COPD compared with each monocomponent used alone. Pharmacoeconomics 23:365–375

Mallia P, Contoli M, Caramori G, Pandit A, Johnston SL, Papi A (2007) Exacerbations of asthma and chronic obstructive pulmonary disease (COPD): focus on virus induced exacerbations. Curr Pharm Des 13:73–97

Martinez FJ, Standiford C, Gay SE (2005) Is it asthma or COPD? The answer determines proper therapy for chronic airflow obstruction. Postgrad Med 117:19–26

Miglino N, Roth M, Tamm M, Borger P (2012) Asthma and COPD – The C/EBP connection. Open Respir Med J 6:1–13

Miravitlles M, Andreu I, Romero Y, Sitjar S, Altés A, Anton E (2012) Difficulties in differential diagnosis of COPD and asthma in primary care. Br J Gen Pract 62: e68–e75

Mishima M (2009) Physiological differences and similarities in asthma and COPD – based on respiratory function testing. Allergol Int 58:333–340

Morice AH, Hochmuth L, Ekelund J, Thorén A, Puterman AS (2008) Comparable long-term safety and efficacy of a novel budesonide/formoterol pressurized metered-dose inhaler versus budesonide/formoterol turbuhaler in adolescents and adults with asthma. Pulm Pharmacol Ther 21:32–39

Mosenifar Z (2009) Differentiating COPD from asthma in clinical practice. Postgrad Med 121:105–112

Nakawah MO, Hawkins C, Barbandi F (2013) Asthma, chronic obstructive pulmonary disease (COPD), and the overlap syndrome. J Am Board Fam Med 26:470–477

Oga T, Tsukino M, Hajiro T, Ikeda A, Koyama H, Mishima M, Chin K, Nishimura K (2010) Multidimensional analyses of long-term clinical courses of asthma and chronic obstructive pulmonary disease. Allergol Int 59:257–265

Pauwels RA (2004) Similarities and differences in asthma and chronic obstructive pulmonary disease exacerbations. Proc Am Thorac Soc 1:73–76

Postma DS, Kerstjens HA (1998) Characteristics of airway hyperresponsiveness in asthma and chronic obstructive pulmonary disease. Am J Respir Crit Care Med 158:S187–S192

Rothe T (2012) COPD and asthma: same but different. Praxis (Bern 1994) 101:233–237

Sciurba FC (2004) Physiologic similarities and differences between COPD and asthma. Chest 126:117S–124S

Silvestri IC, Pereira CA, Rodrigues SC (2008) Comparison of spirometric changes in the response to bronchodilators of patients with asthma or chronic obstructive pulmonary disease. J Bras Pneumol 34:675–682

Sköld CM (2010) Remodeling in asthma and COPD-differences and similarities. Clin Respir J 4(Suppl 1):20–27

Spencer P, Krieger B (2013) The differentiation of chronic obstructive pulmonary disease from asthma: a review of current diagnostic and treatment recommendations. Open Nurs J 7:29–34

Sutherland ER, Martin RJ (2003) Airway inflammation in chronic obstructive pulmonary disease: comparisons with asthma. J Allergy Clin Immunol 112:819–827

Tandon S, Khutarkar A, Ansari S (2013) Asthma diagnosis and treatment – 1008. Is small airways disease a widely prevalent yet underdiagnosed phenotype of asthma and COPD in India? World Allergy Organ J 6(Suppl 1):P8

Tinkelman DG, Price DB, Nordyke RJ, Halbert RJ (2006) Misdiagnosis of COPD and asthma in primary care patients 40 years of age and over. J Asthma 43:75–80

Welte T, Groneberg DA (2006) Asthma and COPD. Exp Toxicol Pathol 57(Suppl 2):35–40

Yawn BP (2009) Differential assessment and management of asthma vs chronic obstructive pulmonary disease. Medscape J Med 11(1):20; Epub 2009

Advs Exp. Medicine, Biology - Neuroscience and Respiration (2016) 21: 39–45
DOI 10.1007/5584_2016_214

Published online: 18 March 2016

Intracellular and Extracellular Cytokines in A549 Cells and THP1 Cells Exposed to Cigarette Smoke

A. Holownia, P. Wielgat, E. Rysiak, and J.J. Braszko

Abstract

Cigarette smoke (CS) activates inflammatory cells and increases cytokine levels producing local and systemic inflammation. To assess changes in intracellular and extracellular cytokine levels we used human epithelial (A549 cells) and monocyte (THP-1) cell lines grown for 24 h in cigarette smoke-conditioned media. Cytokines were assessed using immunostaining/flow cytometry and ELISA assay. In THP1cells, grown in CS-conditioned media, the intracellular interleukins IL-1β, IL-6, and IL-10 increased by more than tenfold, while less significant increases were found in A549 cells. IL-1α and IL-1β, but not IL-6 or IL-10, were increased in the culture media, while IL-2 was raised by about fivefold only in the culture medium of A549 cells. IL-4, IL-6, IL-8, IL-10, IL-12, and tumor necrosis factor alpha were undetectable, while only a slight increase was observed in extracellular IL-17A (by about 60 %) in the medium of A549 cells and by about 115 % in the medium of THP1 cells. The interferon gamma (IFNγ) was increased by about eightfold, but only in the medium of THP1 cells grown with CS. We conclude that IL-1 and INFγ are the key cytokines responsible for pro-inflammatory signaling in epithelial cells and monocytes, respectively, exposed to cigarette smoke.

Keywords

A549 cells • Cell culture • Cigarette smoke • Cytokine • Inflammation • THP1 cells

A. Holownia (✉), P. Wielgat, and J.J. Braszko
Department of Clinical Pharmacology, Medical University of Bialystok, 15a Waszyngtona St., Bialystok, Poland
e-mail: holow_sinai@hotmail.com

E. Rysiak
Department of Drug Chemistry, Medical University of Bialystok, Bialystok, Poland

1 Introduction

Cigarette smoke (CS) is one of the leading causes of death and an important risk factor for systemic and respiratory tract diseases. Chronic CS exposure causes structural and functional changes in the respiratory tract but detailed mechanisms remain obscure. Long-lasting smoking may generate different pathologies including chronic obstructive pulmonary disease (COPD), epithelial cell tumors, cardiovascular disease, but also an increased incidence of asthma and respiratory infections (Chang et al. 2015). Respiratory epithelium is the main target of highly toxic, fresh CS. It has been shown that epithelial integrity and immunity is significantly affected by smoke exposure (Crotty Alexander et al. 2015). CS is clearly detrimental to lung epithelium and it mobilizes and activates alveolar macrophages producing pro-inflammatory mediators, reactive oxygen species, and proteolytic enzymes (Sarma and Ward 2011). On the other hand, several chemicals of CS have both anti-inflammatory and immunosuppressive proprieties (Das et al. 2012; Kalra et al. 2000). Due to the complex nature of CS, diversity of smoking habits, and different experimental models of CS exposure, published data are inconsistent. It has been shown that smoking may suppresses cytokine expression in asthma, but at the same time it evidently worsens lung functions (Tamimi et al. 2012). It seems credible that at early stages circulating monocytes adhere to damaged epithelial cells repeatedly exposed to extremely toxic constituents of CS and migrate into respiratory tissue contributing to morphological and functional changes. We have previously shown that lung epithelial cell line (A549 cells) grown for 24 h in a CS-conditioned culture medium die due to chemical and oxidative stress while human monocyte cells (THP1 cell line) grown in similar CS-saturated medium become activated (Holownia et al. 2015). Moreover, CS toxicity to A549 cells is significantly lower when the cells were co-cultured with THP1 cells sharing a common culture medium. The purpose of the present study was to further explore the same model and investigate the effects of CS on immune response, particularly on intracellular cytokine levels and on cytokine secretion to the culture medium.

2 Methods

2.1 Cell Culture

A549 (ATCC® CCL185™) cells grown in ATCC-formulated F12K medium supplemented with 10 % fetal bovine serum (FBS) and THP1 cells (ATCC® TIB202™) grown in ATCC-formulated RPMI 1640 medium, supplemented with 2-mercaptoethanol to a final concentration of 0.05 mM and with FBS to a final concentration of 10 % were used in this study. Cells were maintained in 37 °C in an incubator in a humidified atmosphere containing 5 % CO_2. For particular experiments cells were plated out onto 6 well plates and were grown in control or smoke conditioned media for 24 h.

2.2 Preparation of CS-Conditioned Media and Cell Treatment

CS-conditioned medium was prepared using full-strength Red Marlboro cigarettes (Phillip Morris; Cracow, Poland) containing 8 mg of tar, 0.6 mg of nicotine, and 9 mg of carbon monoxide per cigarette. To prepare smoke-conditioned media cigarette filters were removed and smoke was passed through culture media (4 cigarettes/ 100 ml of medium) using low pressure vacuum pump. Freshly prepared SC media were diluted with standard media to obtain 30 μM nitrate/ nitrite (colorimetric reaction with Griess reagent) content in each batch. CS-conditioned media were subsequently filtered using 0.22-μm filters and were applied immediately to cell culture. Cells were grown in CS-conditioned media for 24 h and then were tested for intracellular interleukin content. Additionally, levels of interleukins IL-1α, IL-1β, IL-2, IL-4, IL-6, IL-8, IL-10, IL-12, IL-17A, interferon gamma (IFNγ), tumor necrosis factor alpha (TNFα), and granulocyte-macrophage colony-stimulating factor (GM-CSF) were determined in the culture media to quantify cytokine release and leaking.

2.3 Intracellular Interleukin IL-1ß, IL-6, and IL-10

Intracellular interleukins were quantified in Triton X-100 permeabilized A549 or THP1 cells using interleukin-specific primary antibodies, corresponding fluorescent secondary antibodies, and flow cytometry detection. Briefly, cells were washed twice with PBS and counted. 10^6 cells were subsequently permeabilized with Triton® X-100 (1 % in PBS) and rabbit polyclonal antibodies to IL-1β, IL-6, or IL-10 (all from Abcam; Cambridge, UK) were separately added to each sample. After 10 min of incubation at room temperature a secondary, isotype specific fluorescein-bound antibody (Sigma-Aldrich, Poznan, Poland) was added and samples were run on an Epics XL flow cytometer (Coulter Electronics, High Wycombe, UK). Reference samples were prepared using the same isotype, but with unspecific primary antibodies (Abcam; Cambridge, UK). Two thousand total events were collected per sample.

2.4 Extracellular Cytokine Profiles

To quantify cytokines released from the cells to their culture media, the Multi-Analyte Inflammatory Cytokine ELISArray Kits (Qiagen, Manchester, UK) were used. A panel of 12 pro – and anti-inflammatory cytokines including IL-1α, IL-1β, IL-2, IL-4, IL-6, IL-8, IL-10, IL-12, IL-17A, IFNγ, TNFα, and GM-CSF were simultaneously analyzed using a sandwich-based enzyme-linked immunosorbent assay (ELISA) in a microplate, coated with cytokine-specific antibodies. After washing, biotinylated antibodies, avidin-horseradish peroxidase conjugate, and peroxidase substrate was added to produce blue color. Samples were compared with the corresponding positive controls (extracellular cytokines) and with matching data from the CS-conditioned media. Microplate reader (KC junior, BioTek Instruments; Highland Park, VT) was used to read the absorbances at 450/570 nm according to the manufacturer's protocol.

2.5 Statistical Analysis

Statistical analysis was performed with a statistics package-Statistica 6.0 software (Statsoft; Cracow, Poland) using one-way or two-way ANOVA followed by the Bonferroni *post hoc* test for selected pairs of data. Results were expressed as means of 4–6 assays ±SD. A p-value of less than 0.05 was considered statistically significant.

3 Results

Table 1 shows intracellular IL-1β, IL-6, and IL-10 levels (flow cytometry), and pro– and anti-inflammatory cytokines IL-1α, IL-1β, IL-2, IL-4, IL-5, IL-6, IL-8, IL-10, IL-12, IL-13, IL-17A, and GM-CSF profiles in culture media of control A549 and THP1 cells and in culture media of A549 and THP1 cells grown for 24 h in the CS-conditioned medium. In both cell types grown for 24 h in the CS-conditioned medium, intracellular interleukin IL-1β, IL-6 (Fig. 1), and IL-10 were significantly increased. Greater increases in intracellular interleukins were observed in THP1 cells, where IL-1β, IL-6, and IL-10 increased by about 11-fold ($p < 0.01$), 13-fold ($p < 0.01$), and 14-fold ($p < 0.01$), respectively. The corresponding values in A549 cells were about threefold ($p < 0.01$), fivefold, and sevenfold ($p < 0.01$). The highest observed increase was detected in the pleiotropic interleukin - IL-10 in THP1 exposed to CS.

Considering changes in the cytokine profiles in the culture medium, more than half of 12 the cytokines tested were undetectable. Significantly increased levels were found in IL-1α (by more than 2 times, $p < 0.01$) in the culture medium of A549 cells and by more than 4 times ($p < 0.01$) in the culture medium of THP1 cells. Similar, but more pronounced, changes were observed in IL-1ß. Its level increased by more than 3 times ($p < 0.01$) and by more than twice ($p < 0.01$) for A549 and THP1 cells, respectively. IL-2 was increased by about fivefold ($p < 0.01$), but only in the culture medium of A549 cells and was not

Table 1 The effect of cigarette smoke (CS) on intracellular interleukin 1β (IL-1β), interleukin 6 (IL-6) and interleukin 10 (IL-10) and on pro- and anti-inflammatory cytokine profiles (IL-1α, IL-1β, IL-2, IL-4, IL-6, IL-8, IL-10, IL-12, IL-17A, IFNγ, TNFα and GM-CSF) in culture media of A549 cells and THP1 cells grown for 24 h in CS-conditioned medium

Intracellular cytokines (relative units)				
	A549 cells		THP1 cells	
	Control	CS	Control	CS
IL-1β	100 ± 27	330 ± 48**	100 ± 21	1124 ± 311**
IL-6	100 ± 22	533 ± 67**	100 ± 33	1321 ± 433**
IL-10	100 ± 19	664 ± 78**	100 ± 31	1420 ± 632**
Cytokine levels in culture media (relative units)				
IL-1α	100 ± 27	223 ± 29**	100 ± 29	433 ± 77**
IL-1β	100 ± 29	330 ± 42**	100 ± 29	237 ± 65**
IL-2	100 ± 19	520 ± 66**	ND	ND
IL-4	ND	ND	ND	ND
IL-6	ND	ND	ND	ND
IL-8	ND	ND	ND	ND
IL-10	ND	ND	ND	ND
IL-12	ND	ND	ND	ND
IL-17A	100 ± 33	163 ± 29*	100 ± 43	215 ± 41*
IFNγ,	ND	ND	100 ± 41	855 ± 81
TNFα	ND	ND	ND	ND
GM-CSF	100 ± 22	74 ± 41	100 ± 28	143 ± 34

Cytokine levels in control cells are expressed as 100 relative units
IL interleukin, *IFNγ* interferon gamma, *TNFα* tumor necrosis factor alpha, *GM-CSF* granulocyte-macrophage colony-stimulating factor
ND not detectable
*p < 0.05; **p < 0.01 for comparisons with the corresponding control cells

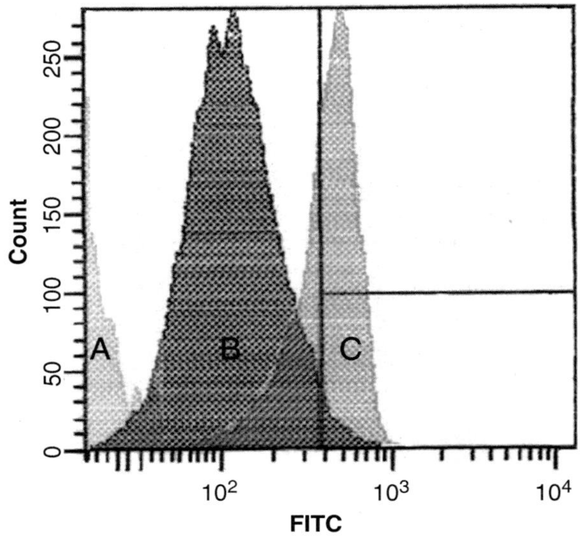

Fig. 1 Histograms of flow cytometry fluorescence of intracellular interleukin 6 (IL-6) in control A549 cells (*B*) and A549 cells grown for 24 h in cigarette smoke-conditioned medium (*C*). Cells were permeabilized, incubated with primary antibody specific to IL-6 and then with fluorescein isothiocyanate (FITC), secondary antibody. Cells were run on epics XL flow cytometer, control gating was set using naïve A549 cells, unspecific primary antibody, and the fluorescent secondary antibody (isotype control; *A*)

detected in THP1 cells. IL-4, IL-6, IL-8, IL-10, and IL-12 were undetectable, while only a slight increase was observed in IL-17A (by about 60 %; $p < 0.05$) in the culture medium of A549 cells and by about 115 % ($p < 0.01$) in the culture medium of THP1 cells. TNFα was not identified, while IFNγ was notably increased (by about eightfold; $p < 0.01$) when compared to control values, but only in the media of THP1 cells grown for 24 h with CS. GM-CSF levels were detectable, but were insignificantly affected by CS exposure.

IL-1β was the only cytokine in A549 and THP1 cells that increased both intracellularly and extracellularly after smoke exposure. Highly elevated levels of intracellular IL-6 and IL-10 upon exposure of A549 and THP1 cells to CS did not result in increased IL-6 and IL-10 levels outside the cells.

4 Discussion

Chronic CS exposure causes structural and functional changes in the respiratory tract, but detailed mechanisms remain elusive. It seems that the primary event in CS cytotoxicity involves epithelial cells, which further activate immune cells to produce inflammation and a variety of time-dependent, morphological and functional alterations. In this study we characterized the epithelial cell interactions with macrophages in a cell culture-based model consisting of human epithelial – A549 cell and monocyte – THP1 cell lines. In our previous study we have shown that the A549 cells and THP1 cells have a contrasting response to CS. The nature and intensity of oxidative and chemical stress induced by CS in the A549 and THP1 cells were different, as CS was significantly more toxic to the A549 cells. Moreover, when both cell types were grown in co-culture sharing a common culture medium, both naïve and CS-pretreated THP1 cells protected the A549 cells against CS toxicity, but also more prone to die (Holownia et al. 2015). In the present report we assessed major cytokine levels inside and outside the A549 and THP1 cells grown in a monoculture, and exposed to CS. We demonstrate that in the A549 and THP1 cells grown for 24 h in the CS-conditioned medium, IL-1β, IL-6 and IL-10 were significantly elevated. Particularly high increases were found in the THP1 cells, where IL-1β and IL-6, and anti-inflammatory IL-10 were increased more than 10 times. In spite of that, both IL-6 and IL-10 were not detected in the culture media, while the pro-inflammatory IL-1β was significantly increased both inside and outside the A549 and THP1 cells.

Inflammatory cytokines play a critical role in coordinating the inflammatory response and are increasingly important targets for therapeutic interventions. Experimental and clinical data show that CS can activate inflammatory cells and stimulate release of inflammatory cytokines (Crotty Alexander et al. 2015). CS increases the levels of IL-1β, IL-6, IL-8, TNFα, and GM-CSF (Zuo et al. 2014; Arnson et al. 2010). In smokers, mononuclear cells produce increased amounts of pro-inflammatory IL-1β, IL-6, and TNFα, and exhibit enhanced response to mitogen-stimuli (Zeidel et al. 2002). Interleukins are classified as pro– and anti-inflammatory, based mostly on their effects on leukocytes. Anti-inflammatory interleukins include IL-4, IL-10, IL-11, and IL-13, while IL-1, IL-6, IL-8, and IFNγ are pro-inflammatory cytokines (Siebert et al. 2015). IL-1 is a potent pro-inflammatory cytokine, which may be activated by tissue damage. It is synthesized as inactive protein and then it is activated by caspase-1 and released outside the cells by several ways, including exocytosis, active transport, cell lysis, and others (Netea et al. 2010). It has been shown that IL-1β is increased in sputum and lavage fluid of smokers (Chung 2006; Ekberg-Jansson et al. 2001). Monocytes isolated from cigarette smokers also produce more IL-1β than do monocytes from non-smokers (Zeidel et al. 2002). Bronchial epithelial cells exposed to CS-conditioned medium also release increase the amount of IL-1β compared with controls (Rusznak et al. 2001). Our present results confirm earlier data and indicate that both macrophages and epithelial cells may increase IL-1β levels and enhance IL-1β export in response to CS. We did not assess the intracellular IL-1α, but increased IL-1α levels were found in the culture media of both cell types exposed to CS. It is possible that IL-1 plays a critical role as inflammatory cytokine in CS exposure.

In the present study, CS significantly increased intracellular IL-10 in the THP1 and A549 cells, but IL-10 remained undetected in the culture media. IL-10 is a pleiotropic cytokine. It downregulates the expression of lymphocyte Th1-derived cytokines and may block pro-inflammatory signaling related to nuclear factor kappa beta (NF-κb) (Sprague and Khalil 2009). It has been found that nicotine may down-regulate IL-10 (Allam et al. 2013). Recently published data also describe the nicotine-induced impairment of IL-10 production by macrophages (Van Zyl-Smit et al. 2014). Our results demonstrate that cells may overexpress or accumulate IL-10 in response to CS smoke, but the protein is not released outside the cells. Further studies are required to explore a detailed causal association between increased cytokine level and its decreased release.

Another extracellular cytokine increased by CS was IL-17A, which is considered as a crucial interleukin that regulates lung immunity and inflammation. Activation of innate cellular sources of IL-17A may mediate the increase in macrophage infiltration of CS-exposed lungs (Bozinovski et al. 2015). Elevated IL-17A has also been observed in sputum collected from exacerbated COPD patients (Roos et al. 2015). Recent knockout studies indicate that IL-17A is not involved in CS-induced loss of lung functions, but rather contributes to normal lung homeostasis (Voss et al. 2015). In our experimental model, IL-17A was increased by about twofold, which is similar to recently published clinical data in COPD patients (Montalbano et al. 2015). In that study, expression of IL-17A in the epithelial cells of distal airways positively correlated with the total packs per year in COPD patients, which might influence the rate of apoptosis and proliferation.

In the present study, INFγ increased in THP1 cells, but not in A549 cells, exposed to CS. An increased level of INFγ in inflammatory cells has been described in several models of CS toxicity (Ahn and Aggarwal 2005), which can reflect the activation of inflammatory pathways. On the other hand, CS did not induce INFγ in our epithelia-derived cells, indicating different regulatory mechanisms.

In conclusion, the present findings indicate that damaged and chemically stimulated epithelial cells exposed to CS activate internal and external inflammatory response, cytokine secretion, and possibly stimulate inflammatory cells. IL-1 seems to play a major role in mediating the CS effects in epithelial cells, while the INFγ−related signaling may be important in macrophages.

Conflicts of Interest The authors had no conflicts of interest to declare in relation to this article.

References

Ahn KS, Aggarwal BB (2005) Transcription factor NF-kappaB: a sensor for smoke and stress signals. Ann N Y Acad Sci 1056:218–233

Allam E, Delacruz K, Ghoneima A, Sun J, Windsor LJ (2013) Effects of tobacco on cytokine expression from human endothelial cells. Oral Dis 19:660–665

Arnson Y, Shoenfeld Y, Amital H (2010) Effects of tobacco smoke on immunity, inflammation and autoimmunity. J Autoimmun 34:258–265

Bozinovski S, Seow HJ, Chan SP, Anthony D, McQualter J, Hansen M, Jenkins BJ, Anderson GP, Vlahos R (2015) Innate cellular sources of interleukin-17A regulate macrophage accumulation in cigarette-smoke-induced lung inflammation in mice. Clin Sci (Lond) 129:785–796

Chang CM, Corey CG, Rostron BL, Apelberg BJ (2015) Systematic review of cigar smoking and all cause and smoking related mortality. BMC Public Health 15:390. doi:10.1186/s12889

Chung KF (2006) Cytokines as targets in chronic obstructive pulmonary disease. Curr Drug Targets 7:675–681

Crotty Alexander LE, Shin S, Hwang JH (2015) Inflammatory diseases of the lung induced by conventional cigarette smoke: a review. Chest 148:1307–1322

Das S, Chakraborty SP, Roy S, Roy S (2012) Nicotine induced pro-oxidant and antioxidant imbalance in rat lymphocytes: in vivo dose and time dependent approaches. Toxicol Mech Methods 22:711–720

Ekberg-Jansson A, Bake B, Andersson B, Skoogh BE, Löfdahl CG (2001) Respiratory symptoms relate to physiological changes and inflammatory markers reflecting central but not peripheral airways. A study in 60-year-old 'healthy' smokers and never-smokers. Respir Med 95:40–47

Holownia A, Wielgat P, Kwolek A, Jackowski K, Braszko JJ (2015) Crosstalk between co-cultured A549 cells and THP1 cells exposed to cigarette smoke. Adv Exp Med Biol 858:47–55

Kalra R, Singh SP, Savage SM, Finch GL, Sopori ML (2000) Effects of cigarette smoke on immune response: chronic exposure to cigarette smoke impairs antigen-mediated signaling in T cells and depletes IP3-sensitive Ca^{2+} stores. J Pharmacol Exp Ther 293:166–171

Montalbano AM, Riccobono L, Siena L, Chiappara G, Di Sano C, Anzalone G, Gagliardo R, Ricciardolo FL, Sorbello V, Pipitone L, Vitulo P, Profita M (2015) Cigarette smoke affects IL-17A, IL-17F and IL-17 receptor expression in the lung tissue: ex vivo and in vitro studies. Cytokine 76:391–402

Netea MG, Simon A, van de Veerdonk F, Kullberg BJ, Van der Meer JW, Joosten LA (2010) IL-beta processing in host defense: beyond the inflammasomes. PLoS Pathog 6:e1000661. doi:10.1371/journal.ppat.1000661

Roos AB, Sethi S, Nikota J, Wrona CT, Dorrington MG, Sandén C, Bauer CM, Shen P, Bowdish D, Stevenson CS, Erjefält JS, Stampfli MR (2015) IL-17A and the promotion of neutrophilia in acute exacerbation of chronic obstructive pulmonary disease. Am J Respir Crit Care Med 192:428–437

Rusznak C, Sapsford RJ, Devalia JL, Shah SS, Hewitt EL, Lamont AG, Davies RJ, Lozewicz S (2001) Interaction of cigarette smoke and house dust mite allergens on inflammatory mediator release from primary cultures of human bronchial epithelial cells. Clin Exp Allergy 3:226–238

Sarma JV, Ward PA (2011) Oxidants and redox signaling in acute lung injury. Compr Physiol 1:1365–1381

Siebert S, Tsoukas A, Robertson J, McInnes I (2015) Cytokines as therapeutic targets in rheumatoid arthritis and other inflammatory diseases. Pharmacol Rev 67:280–309

Sprague AH, Khalil RA (2009) Inflammatory cytokines in vascular dysfunction and vascular disease. Biochem Pharmacol 78:539–552

Tamimi A, Serdarevic D, Hanania NA (2012) The effects of cigarette smoke on airway inflammation in asthma and COPD: therapeutic implications. Respir Med 106:319–328

Van Zyl-Smit RN, Binder A, Meldau R, Semple PL, Evans A, Smith P, Bateman ED, Dheda K (2014) Cigarette smoke impairs cytokine responses and BCG containment in alveolar macrophages. Thorax 69:363–370

Voss M, Wolf L, Kamyschnikow A, Wonnenberg B, Honecker A, Herr C, Lepper PM, Wegmann M, Menger MD, Bals R, Beisswenger C (2015) Il-17A contributes to maintenance of pulmonary homeostasis in a murine model of cigarette smoke-induced emphysema. Am J Physiol Lung Cell Mol Physiol 309:188–195

Zeidel A, Beilin B, Yardeni I, Mayburd E, Smirnov G, Bessler H (2002) Immune response in asymptomatic smokers. Acta Anaesthesiol Scand 46:959–964

Zuo L, He F, Sergakis GG, Koozehchian MS, Stimpfl JN, Rong Y, Diaz PT, Best TM (2014) Interrelated role of cigarette smoking, oxidative stress, and immune response in COPD and corresponding treatments. Am J Physiol Lung Cell Mol Physiol 307:205–218

Advs Exp. Medicine, Biology - Neuroscience and Respiration (2016) 21: 47–53
DOI 10.1007/5584_2016_213

Published online: 18 March 2016

Frequency of Rare Alpha-1 Antitrypsin Variants in Polish Patients with Chronic Respiratory Disorders

K. Duk, A. Zdral, B. Szumna, A. Roży, and J. Chorostowska-Wynimko

Abstract

The *SERPINA1* gene encoding the alpha-1 antitrypsin (A1AT) protein is highly polymorphic. It is known that, apart from the most prevalent PI*S and PI*Z A1AT deficiency variants, other so-called rare variants also predispose individuals to severe chronic respiratory disorders such as emphysema and chronic obstructive pulmonary disease. Our aim was to assess the frequencies of common and rare *SERPINA1* mutations in a group of 1033 Polish patients referred for A1AT deficiency diagnostics due to chronic respiratory disorders in the period of January 2014–September 2015. All blood samples were analyzed according to the routine diagnostic protocol, including A1AT serum concentration assessment by nephelometry and immune isoelectric focusing, followed by PCR genotyping and direct sequencing when necessary. A total of 890 out of the 1033 samples (86 %) carried the normal PI*MM genotype, whereas, in 143 samples (14 %), at least one A1AT deficiency variant was detected. In 132 subjects, PI*S (2.1 %) and PI*Z (10.8 %) common deficiency alleles were identified, yielding frequencies of 0.011 and 0.062, respectively. Rare *SERPINA1* variants were detected in nine patients: PI*F (c.739C>T) (n = 5) and PI*I (c.187C>T) (n = 4). Samples from the patients with an A1AT serum concentration below 120 mg/dl and presenting a PI*MM-like phenotypic pattern were retrospectively analyzed by direct sequencing for rare *SERPINA1* mutations, revealing a PI*M2$_{\text{Obernburg}}$ (c.514G>T) mutation in one patient and a non-pathogenic mutation (c.922G>T) in another. We conclude that the deficiency PI*Z A1AT allele is considerably more common in patients with chronic respiratory disorders than in the general Polish population. The prevalence of the PI*F allele seems higher than in other European studies.

K. Duk, A. Zdral, B. Szumna, A. Roży, and J. Chorostowska-Wynimko (✉)
Department of Genetics and Clinical Immunology, National Institute of Tuberculosis and Lung Diseases, 26 Płocka St., 01-138 Warsaw, Poland
e-mail: j.chorostowska@igichp.edu.pl

Keywords
Alpha-1 antitrypsin • Diagnosis of deficiency • Hereditary disorders • SERPINA1 • Antitrypsin variants

1 Introduction

Alpha-1 antitrypsin deficiency (A1ATD), one of the three most common inherited disorders in Caucasians, is an autosomal codominant hereditary disorder predisposing individuals to pulmonary and liver disorders. The pathophysiology of A1ATD is directly related to a mutation of the *SERPINA1* gene, located on the long arm of chromosome 14 (14q31–32.3). *SERPINA1* encodes alpha-1 antitrypsin (A1AT), a small protein (394 amino acids) that is produced mostly by hepatocytes, but also, at lower concentrations, by respiratory and intestinal epithelial cells, neutrophils, and macrophages (Ferrarotti et al. 2007).

A1AT is characterised by a broad spectrum of activities. As an acute-phase protein and a major serine protease inhibitor with affinity to neutrophil elastase (ELANE2) and proteinase 3 (PR3), it prevents uncontrolled proteolysis in the connective tissue of the lower respiratory tract, particularly during acute and chronic inflammatory events. In addition, A1AT exhibits exceptional immunomodulatory properties that are crucial for physiological regulation and termination of the inflammatory response (Janciauskiene et al. 2011a).

SERPINA1 is a highly polymorphic gene. More than 125 mutations have been identified, including at least 60 encoding clinically meaningful A1AT protein variants (Popławska et al. 2013). The two most common deficiency alleles, PI*S (c.863A>T; p.Glu264Val) and PI*Z (c.1096G>A; p.Glu342Lys), are relatively prevalent in the Caucasian population (5–10 % and 1–3 %, respectively). PI*S is widely spread in Europe, with its frequency increasing from northeast to southwest and peaking on the Iberian Peninsula (more than 70 cases per 1000 individuals: 0.07). The PI*Z allele is more prevalent in north-eastern Europe, Scandinavia, and the Baltic region (0.02–0.04) (Fernandes da Silva 2014).

The most severe A1ATD results from PI*Z homozygosity (PI*ZZ) or is due to the null variant (PI*null, null; PI*Z null) (Fregonese et al. 2008). There are also a considerable number of so-called rare *SERPINA1* mutations, some characterised by equally severe clinical consequences. The PI*Z and PI*S alleles are easily identified by standard diagnostic geno- or phenotyping, whereas rare variants require more sophisticated methods such as gene sequencing. In some, the risk of misdiagnosis is particularly high due to either an unspecific phenotypic pattern or a normal A1AT serum concentration. For example, PI*Mmalton (p.Phe51del/p.Phe52del; c.227_229delTCT) is characterized by a lower A1AT serum concentration and a concomitantly normal PI*MM-like isoelectric pattern (Belmonte et al. 2015). Meanwhile, its clinical respiratory and liver presentation might be very similar to severe PI*ZZ deficiency. Intrahepatic polymers of newly synthesized misfolded Mmalton A1AT protein affect cellular homeostasis, induce oxidative stress and damage hepatocytes, increasing the risk of liver damage. Additionally, owing to retention of the polymer in the liver, the A1AT concentration in the blood and tissues is reduced, with a corresponding increased risk of lung disease (Janciauskiene et al. 2011b). According to recent data, the prevalence of PI*Mmalton is relatively high in the Mediterranean region. In Italy, rare variants, including PI*Mmalton, PI*Mprocida (c.194T>C; p.Leu41Pro), PI*P_{Lowell} (c.839A>T, p. Asp280Val), and PI*I (c.187C>T; p.Arg39Cys) (frequency 0.077), are more prevalent than PI*Z (0.0013) (Ferrarotti et al. 2005). Likewise, the PI*I allele accounts for as much as 34 % of all rare

A1AT variants in the Spanish population (Rodriguez-Frias et al. 2012).

In the Polish population, the frequencies of the two most common deficiency alleles, PI*S and PI*Z, have been estimated to be 0.0145 and 0.0109, respectively (Chorostowska-Wynimko et al. 2012; Kaczor et al. 2007), but there are no data regarding rare variants. Thus, the aim of this study was to evaluate the frequency of rare A1AT variants in a group of Polish patients with chronic respiratory disorders referred for A1ATD diagnostics.

2 Methods

A local Ethics Committee approved the study protocol (permit no. KB-170/2011). Samples from 1033 patients (566 females and 467 males, aged 54 ± 19 years) with chronic respiratory disorders (emphysema, bronchial asthma, chronic obstructive pulmonary disease (COPD), bronchiectasis, and chronic bronchitis) referred for A1ATD diagnostics were included in the study. First, they were analyzed in accordance with the diagnostic algorithm worked out by the Polish Respiratory Society (Fig. 1). Later, a group of 100 patients was selected according to a protocol for rare variants with preset criteria, namely, a normal (PI*MM) phenotype and an A1AT serum concentration below 120 mg/dl, to undergo Sanger sequencing of four exons (II, III, IV, V) of the *SERPINA1* gene.

Blood samples were collected on ethylenediaminetetraacetic acid (EDTA) as an anticoagulant and a clot activator. All samples were stored at −70 °C until further analysis. Dried blood spots, EDTA-treated, were collected on a filter paper (Whatman #903), as described previously (Chorostowska-Wynimko et al. 2012).

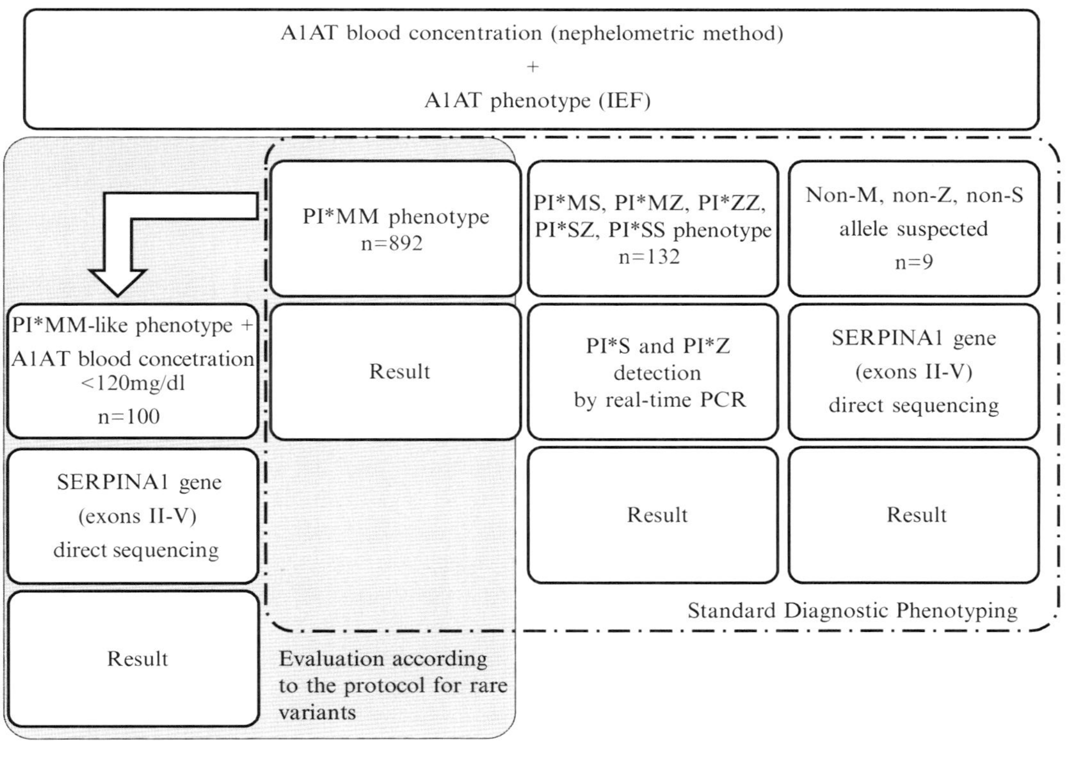

Fig. 1 Schematic representation of the alpha-1 antitrypsin deficiency diagnostic strategy applied in the study

2.1 Genotyping

A1AT blood concentration was measured by the nephelometric method (Immage 800; Beckman Coulter; Brea, CA). A1AT phenotyping was carried out using the Sebia HydraSys electrophoresis platform and the Hydragel 18 A1AT Isofocusing Kit (Sebia; Lisses, France) by isoelectric focusing (IEF) (Zerimech et al. 2008). DNA extraction from dried blood spots and the cellular fraction was performed using a commercially available kit: Extract-N-Amp Blood PCR Kit (Sigma-Aldrich; Poznan, Poland), GeneMATRIX or Bio-Trace DNA Purification Kit (EURx Ltd.; Gdansk, Poland). Genetic material present in the eluate was directly used for A1AT genotyping without the need for DNA purification.

Identification of the two most common mutations of the *SERPINA1* gene (PI*Z, PI*S) was performed in a single reaction by real-time PCR in a LightCycler 480 II instrument (Roche Diagnostics Ltd.; Rotkreuz, Switzerland) using hydrolyzing probes coupled with fluorescent dyes (VIC or FAM) complementary to the mutant variants (PI*S or PI*Z). Primer and probe sequences and PCR reaction conditions were previously described (Struniawski et al. 2013). Rare A1AT variants were confirmed by direct sequencing. Sequence analysis of A1AT exons II–V was performed with the 16-capillary 3130xl Genetic Analyzer (Genomed; Warsaw, Poland).

3 Results

3.1 Evaluation According to the Standard Diagnostic Protocol for A1ATD

The routine A1ATD diagnostic algorithm, as proposed by the Polish Respiratory Society guidelines, involves a stepwise approach including the assessment of the A1AT blood concentration and A1AT phenotyping (Fig. 1). If a deficient A1AT protein variant is detected, the DNA is re-evaluated by genotyping or direct sequencing prior to the final diagnosis.

A total of 1033 samples from patients with chronic respiratory disorders referred for routine A1AT diagnostics in the period of January 2014–September 2015 were evaluated according to the standard protocol. The group comprised 467 males (45 %) and 566 females (55 %), with a mean age of 56 $\pm$ 19 years. The mean A1AT concentration assessed by nephelometry was 151 $\pm$ 48 mg/dl. Immunophenotyping confirmed a normal A1AT PI*MM phenotype in 892 of the 1033 samples (86 %). Interestingly, in 110 individuals with the PI*MM phenotype, the A1AT concentration was below 120 mg/ml (mean 111 $\pm$ 19 mg/dl), the diagnostic cut-off level suggested by the Polish guidelines.

In 141 samples (14 %), at least one A1AT deficiency variant was detected. In 132 subjects, PI*S (2.1 %) or PI*Z (10.8 %) common deficiency alleles were identified, yielding frequencies of 0.011 and 0.062, respectively. The homozygous PI*ZZ genotype was detected in 16 individuals (1.5 %), PI*MZ in 94 (9 %), PI*MS in 20 (1.9 %), and PI*SZ in 2 (0.2 %). In addition, five individuals with the PI*FM (c.739C > T; p.Arg223Cys) (0.5 %) and four with the PI*IM genotype (c.187C>T; p.Arg39Cys) (0.4 %) were diagnosed. Isoelectric patterns of rare A1AT gene mutations were confirmed by A1AT sequencing.

3.2 Evaluation According to the Protocol for Rare Variants

A total of 100 of the 892 patients (11 %) were selected according to the following preset criteria: a normal (PI*MM) phenotype and an A1AT serum concentration below 120 mg/dl. Samples were re-examined by Sanger sequencing to assess the presence of rare mutations in four exons (II, III, IV, and V) of the *SERPINA1* gene.

There were 61 male and 39 female patients in the subgroup; the average A1AT serum concentration was 107 $\pm$ 14 mg/dl. Direct sequencing revealed one subject with the PI*$M2_{Obernburg}$ variant (c.514G > T; p.Gly148Trp) and another

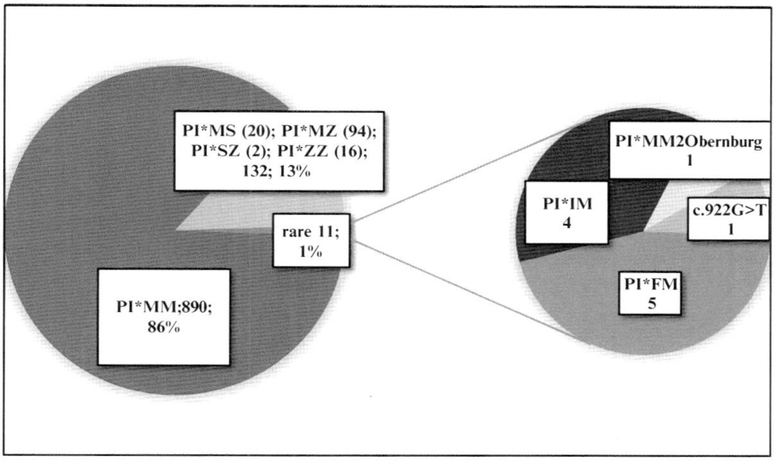

Fig. 2 Alpha-1 antitrypsin genotypes observed in the study

with the presumably non-pathogenic mutation (c.922G > T) (Fig. 2).

3.3 Clinical Characteristics

Patients with respiratory diseases and rare A1AT mutations suffered predominantly from COPD (n = 6, 55 %) (PI*FM n = 5; PI*IM n = 1), while the emphysematous phenotype was also frequent (n = 2, 18 %) (PI*MM2$_{Obernburg}$ n = 1; PI*IM n = 1). However, granulomatosis with polyangiitis (formerly Wegener's disease) was diagnosed in two subjects (18 %) and interstitial lung disease in one (9 %).

4 Discussion

This study represents the very first attempt to assess the frequency of rare *SERPINA1* gene mutations in the Polish population. The study group consisted of 1033 subjects with chronic respiratory disorders referred for A1ATD diagnostics. Owing to the bias caused by the inclusion criteria and a relatively small number of examined subjects, the presented data should be considered as preliminary. Nevertheless, the calculated frequencies of the common deficiency alleles PI*S and PI*Z are 0.011 and 0.062, respectively, which are higher than previously demonstrated for the general Polish population (Chorostowska-Wynimko et al. 2012), supporting the need for diagnostic screening for A1AT deficiency in patients with chronic respiratory disorders. We also detected a relatively large number of subjects with rare variants, including patients with PI*F, PI*I and PI*M2$_{Obernburg}$. The calculated PI*F prevalence (0.002) is higher than that shown for other European populations. There was only one carrier of the PI*F allele found among 3511 Spanish patients and two in a Swiss group of 1399 subjects (Rodriguez-Frias et al. 2012). It is worth emphasising that, while there are limited data regarding the prevalence of rare mutations in Europe, there are also considerable regional differences throughout the continent. A retrospective analysis demonstrated a relatively high frequency of PI*Mmalton among the non-S and non-Z deficiency variants in the Mediterranean population: 20 % in Spain, 60 % in Tunisia, and 35 % in Italy, but only 8 % in Switzerland. Likewise, the PI*I allele accounts for 34 % of rare A1AT variants in Spain and 28 % in Switzerland, but only 5 % in Italy (Rodriguez-Frias et al. 2012). It has been suggested that rare variants may be more common in areas with a low prevalence of the main deficient variants (PI*S, PI*Z), as shown for Italy (Ferrarotti et al. 2005), where rare variants are almost 60 times more common than the PI*Z allele. In

Ireland, the relatively high prevalence rates of PI*Z (allele frequency 0.094) and PI*S (allele frequency 0.052) variants are accompanied by a small number of rare mutations (0.007), such as PI*I, PI*F, PI*V, PI*Mmalton, and PI*Zbristol (Carroll et al. 2011; Alpha One Foundation Annual Report 2012).

Similar to our study, the majority of epidemiological data on *SERPINA1* rare mutations have been generated from patients with chronic respiratory disorders (Ferrarotti et al. 2008). As a rule, large-scale studies addressing the epidemiology of A1AT in the representative general population have assessed only the two most common alleles, PI*S and PI*Z (Rodríguez-Frías et al. 2011). Consequently, currently available reliable records on A1ATD prevalence only refer to the PI*S and PI*Z mutations (de Serres and Blanco 2012).

It should also be emphasised that most standard diagnostic protocols preclude the detection of a considerable number of rare mutations, mainly those encoding qualitative deficiencies, such as PI*F, but also those with a misleading isoelectric pattern, such as PI*Mmalton. This implies the need for a more individualised approach to diagnostics than proposed by commonly implemented algorithms, or alternatively indicates the need for new, more informative diagnostic tests. Gene sequencing is the only definitive way to identify *SERPINA1* mutations, yet it is time-consuming and relatively expensive, and thus impractical for diagnostic screening (Struniawski et al. 2008). Therefore, multiplex PCR genotyping identifying not only S- and Z- but also a number of rare mutations could be a very useful alternative. Importantly, the assessment of clinical symptoms would not help as an identifying criterion. In addition to typical pulmonary (emphysema, COPD, and bronchial asthma) and liver disorders (cirrhosis, neonatal hepatitis, and hepatocellular carcinoma) (Curiel et al. 1989), A1ATD has been consistently associated with other diseases such as vasculitis and granulomatosis with polyangiitis (Chorostowska-Wynimko et al. 2013). This finding is well represented in our group. In addition to emphysema and COPD, there also were patients suffering from granulomatosis with polyangiitis and interstitial lung fibrosis.

In conclusion, the current study presents the very first data on the prevalence of rare mutations in a Polish population of patients with chronic respiratory disorders. Interestingly, the PI*F allele seems to be more common than in other European studies. To confirm its higher prevalence in our region, more research is needed.

Acknowledgments This study was performed as part of the scientific project: Dissemination and optimization of alpha-1 antitrypsin deficiency diagnostic algorithm in patients with chronic lung diseases (theme 5/4) of the National Institute of Tuberculosis and Lung Diseases, Warsaw in Poland. The authors are deeply indebted to the Polish Foundation for Patients with alpha-1 antitrypsin deficiency for their financial support of this research.

Conflicts of Interest The authors declare no conflicts of interest in relation to this article.

References

Alpha One Foundation Annual Report (2012) Available from: http://www.alpha1.ie/info-centre/annual-reports. Accessed 24 Dec 2015

Belmonte I, Montoto L, Miravitlles M, Barrecheguren M, Esquinas C, Rodríguez E, Giralt M, Rodríguez-Frías F (2015) Rapid detection of Mmalton α1-antitrypsin deficiency allele by real-time PCR and melting curves in whole blood, serum and dried blood spot samples. Clin Chem Lab Med 54:1434–6621. doi:10.1515/cclm-2015-0297

Carroll TP, O'Connor CA, Floyd O, McPartlin J, Kelleher DP, O'Brien G, Dimitrov BD, Morris VB, Taggart CC, McElvaney NG (2011) The prevalence of alpha-1 antitrypsin deficiency in Ireland. Respir Res 12:91. doi:10.1186/1465-9921-12-91

Chorostowska-Wynimko J, Struniawski R, Popławska B, Borszewska-Kornacka M (2012) The incidence of alpha-1-antitrypsin (A1AT) deficiency alleles in population of Central Poland – preliminary results from newborn screening. Pneumonol Alergol Pol 80:450–453

Chorostowska-Wynimko J, Gawryluk D, Struniawski R, Popławska B, Fijołek J (2013) Incidence of alpha-1 antitrypsin Z and S alleles in patients with granulomatosis with polyangiitis – pilot study. Pneumonol Alergol Pol 81:319–322

Curiel DT, Holmes MD, Okayama H, Brantly ML, Vogelmeier C, Travis WD (1989) Molecular basis of the liver and lung disease associated with the alpha

1-antitrypsin deficiency allele Mmalton. J Biol Chem 264:13938–13945

de Serres FJ, Blanco I (2012) Prevalence of α1-antitrypsin deficiency alleles PI*S and PI*Z worldwide and effective screening for each of the five phenotypic classes PI*MS, PI*MZ, PI*SS, PI*SZ, and PI*ZZ: a comprehensive review. Ther Adv Respir Dis 6:277–295

Fernandes da Silva DI (2014) Alpha-1-Antitrypsin deficiency, exploring the role of SERPINA1 rare variants and searching for genetic modifiers of associated diseases (Granulomatosis with Polyangiitis). Available from: https://repositorio-aberto.up.pt/bitstream/10216/77090/2/104607.pdf. Accessed 24 Dec 2015

Ferrarotti I, Baccheschi J, Zorzetto M, Tinelli C, Corda L, Balbi B, Campo I, Pozzi E, Faa G, Coni P, Massi G, Stella G, Luisetti M (2005) Prevalence and phenotype of subjects carrying rare variants in the Italian registry for alpha1-antitrypsin deficiency. J Med Genet 42:282–287

Ferrarotti I, Scabini R, Campo I, Ottaviani S, Zorzetto M, Gorrini M, Luisetti M (2007) Laboratory diagnosis of alpha1-antitrypsin deficiency. Transl Res 150:267–274

Ferrarotti I, Gorrini M, Scabini R, Ottaviani S, Mazzola P, Campo I, Zorzetto M, Luisetti M (2008) Secondary outputs of alpha1-antitrypsin deficiency targeted detection programme. Respir Med 102:354–358

Fregonese L, Stolk J, Frants RR, Veldhuisen B (2008) Alpha-1 antitrypsin Null mutations and severity of emphysema. Respir Med 102:876–884

Janciauskiene SM, Bals R, Koczulla R, Vogelmeier C, Köhnlein T, Welte T (2011a) The discovery of α1-antitrypsin and its role in health and disease. Respir Med 105:1129–1139

Janciauskiene S, Ferrarotti I, Laenger F, Jonigk D, Luisetti M (2011b) Clinic card for: α-1-antitrypsin deficiency. Eur J Hum Genet 19:1–3

Kaczor M, Sanak M, Libura-Twardowska M, Szczeklik A (2007) The prevalence of alpha1-antitrypsin deficiency in a representative population sample from Poland. Respir Med 101:2520–2525

Popławska B, Janciauskiene S, Chorostowska-Wynimko J (2013) Genetic variants of alpha-1 antitrypsin: classification and clinical implications. Pneumonol Alergol Pol 81(1):45–54 (Article in Polish)

Rodríguez-Frías F, Vila-Auli B, Homs-Riba M, Vidal-Pla R, Calpe-Calpe JL, Jardi-Margalef R (2011) Diagnosis of alpha-1 antitrypsin deficiency: limitations of rapid diagnostic laboratory tests. Arch Bronconeumol 47:415–417

Rodriguez-Frias F, Miravitlles M, Vidal R, Camos S, Jardi R (2012) Rare alpha-1-antitrypsin variants: are they really so rare? Ther Adv Respir Dis 6:79–85

Struniawski R, Szpechciński A, Chorostowska-Wynimko J (2008) Molecular diagnostics of alpha-1-antitrypsin deficiency in clinical practice. Pneumonol Alergol Pol 76:253–264 (Article in Polish)

Struniawski R, Szpechcinski A, Popławska B (2013) Rapid DNA extraction protocol for the alpha-1 antitrypsin deficiency detection from dried blood spots by real-time PCR. Adv Exp Med Biol 756:29–37

Zerimech F, Hennache G, Bellon F, Barouh G, Lafitte J (2008) Evaluation of a new Sebia isoelectrofocusing kit for alpha 1-antitrypsin phenotyping with the Hydrasys System. Clin Chem Lab Med 46:260–263

Advs Exp. Medicine, Biology - Neuroscience and Respiration (2016) 21: 55–62
DOI 10.1007/5584_2015_208

Published online: 29 January 2016

Influence of Body Shape Composition on Respiratory Function in Adult Women

Z. Czapla, A. John, A. Szwed, T. Hanć, M. Durda, J. Ratajczak, and E. Barłóg

Abstract

The purpose of the study was to evaluate the influence of body size and shape, and of fat distribution on respiratory functions in adult women. The sample consisted of 107 women aged 17–82 years. Height, weight, chest, waist and hip circumferences, abdominal, and subscapular and triceps skinfolds were examined. The BMI and WHR were calculated. Forced vital capacity (FVC), forced expiratory volume in 1 s (FEV_1), peak expiratory flow (PEF), maximal expiratory flow (MEF_{75}), forced expiratory time (FET) were used as measures of respiratory function. Positive correlations were found between z-scores of height and VC, FEV_1, FVC, between WHR and VC, and between circumference and FET. Negative correlations were found between z-scores of BMI and VC, FEV_1, between subscapular skinfold and VC, FEV_1, FVC and between abdominal skinfold and VC. Stepwise multiple regression analysis showed that traits of body size and shape mostly influenced VC (32 %) and FVC (31 %). Body height and WHR significantly affected VC, while height and subscapular skinfold affected FVC. A significant association between fat distribution described by BMI, WHR, and skinfold thickness and respiratory parameters was observed. These results confirm a complex effect of body size and shape, and of fat distribution on respiratory function.

Keywords

Fat distribution • Forced expiratory volume • Respiratory function • Spirometry • Vital capacity • Waist-to-hip ratio

Z. Czapla (✉), A. John, A. Szwed, T. Hanć, M. Durda, J. Ratajczak, and E. Barłóg
Department of Human Biological Development, Institute of Anthropology, Faculty of Biology, Adam Mickiewicz University, 89 Umultowska St, 61-614 Poznan, Poland
e-mail: czapla@amu.edu.pl

1 Introduction

Respiratory function is essential for maintaining homeostasis. Respiratory system disorders reduce quality of life and may contribute to the development of diseases of other systems, e.g., cardiovascular, and increase the risk of cancer (Hole et al. 1996).

A multitude of different factors affect the functioning of the respiratory system, including genetic (Sandford et al. 2001) and environmental factors, for instance air pollution (Gauderman et al. 2000) or black carbon (Suglia et al. 2008). Increasing more attention is focused on the influence of overweight and obesity on respiration (Chen et al. 2007). Body size disturbances lead to negative modifications in respiratory system functioning (Lessard et al. 2011; Li et al. 2003). Previous studies have shown that obesity is connected with chronic obstructive pulmonary disease, aspiration pneumonia, asthma, obstructive sleep apnea, and pulmonary embolism (McClean et al. 2008). In this context it is worth noting that Saliman et al. (2008) have demonstrated the occurrence of extremely obese patients with normal spirometry and lung volumes.

The majority of studies, however, show a negative association of overweight and obesity with forced vital capacity (FVC) and forced expiratory volume in 1 second (FEV_1) (Melo et al. 2014). Babb et al. (2008) have demonstrated a negative effect of BMI, percentage of body fat, and regional obesity (anterior subcutaneous abdominal fat and visceral fat) on respiratory function in obese men and women. In contrast, other reports show that not so much BMI as body fat and fat-free mass have an impact on respiration (Cotes et al. 2001; Lazarus et al. 1998). Some reports suggest that accumulation of fat in specific regions may be essential for respiratory deterioration. For instance, Collins et al. (1995) have shown that upper body fat, assessed from the biceps skinfold, significantly affects respiration. Also, impact on FVC and FEV_1 of fat accumulation expressed as waist circumference has been demonstrated (Chen et al. 2007). Further, Harik-Khan et al. (2001) have found that reductions of FVC are significantly associated with waist-to-hip ratio (WHR). The difference in fat distribution between men and women should be considered in such studies as it may confound the interpretation of results. The purpose of the present study was to evaluate the influence of morphological traits connected with body size and shape, and fat distribution on respiratory function in adult women.

2 Methods

The study was performed in accordance with the ethical standards laid down in the Declaration of Helsinki for Human Research and with the national law. A total of 145 female patients from the Zdrovita Medical Center of Pulmonary Diseases in the town of Nowy Tomyśl in Poland were initially enrolled into the study. The main inclusion criterion was a pseudo-Tiffeneau index (FEV_1/FVC) >0.7 and FVC >80 %, which excludes obstructive and restrictive lung conditions. The criterion was met by 107 women (mean age of 51 ± 16, range of 17–82 years) who entered further study procedures. The following body size and shape characteristics were assessed: height, weight, chest, and waist and hip circumferences. The skinfold thickness was measured in the triceps, subscapular, and abdominal regions. All measurements were taken according to the standardized procedures described by Martin and Saller (1957). Body height was measured with an anthropometer with accuracy of ±1 cm. Body weight accuracy was of ±100 g. An anthropometric tape was used to measure the body circumferences to the nearest 0.5 cm. Skinfolds were measured using a calliper (Holtain Ltd.; Crymych, UK) with accuracy of ±1 mm. BMI was calculated from height and body weight (kg/m^2). Spirometry was carried out in a sitting position (MicroLab ML 3500; Micro Medical Ltd.; Newcastle upon Tyne, UK). The following respiratory variables were taken into account: vital capacity (VC), forced vital capacity (FVC), forced expiratory volume during the first second of expiration (FEV_1), peak expiratory flow (PEF), maximal expiratory flow at 75 % of FVC (MEF_{75}), and forced expiratory time (FET). Spirometer accuracy was ±3 % in accordance with the ATS standards of 1994 in terms of flow and volume. Beside anthropometric and spirometric tests, participants filled out a questionnaire consisting of questions about age, education, and air pollution in the work environment.

Statistical elaboration consisted of the assessment of associations between the traits measured with Pearson's correlation coefficient. Stepwise multiple regression was used to assess the influence of body size and shape features on respiratory variables. The database was created in Microsoft Excel 2007, and all calculations were conducted with Statistica 9.1. software. Results were considered statistically significant at $p < 0.05$.

3 Results

The age of women strongly correlated with body size and shape. A significant positive correlation with age was present for most of the morphological traits studied, except the triceps skinfold. Body height and WHR correlated negatively with age (height: r =−0.50 and WHR r =−0.54). A negative correlation was present between age and all functional respiratory variables, except FET (Table 1). These findings indicate that advancing age of study participants fostered weight gain and also deterioration of respiratory function. Due to this strong dependence on age of the traits studied, data were adjusted for age in further statistical elaboration.

We examined correlations between z-scores of body size and shape features and z-scores of respiratory variables. A significant positive correlation was found between body height, on one side, and VC, FEV_1, and FVC on the other side (r = 0.524; r = 0.481; and r = 0.538, respectively), and also between WHR and VC (r = 0.250). The z-score of chest circumference correlated positively with that of FET (r = 0.204), while z-score of BMI correlated negatively those of VC and FEV_1 (r = −0.204 and r = −0.220, respectively). A negative correlation was obtained between z-score of subscapular skinfold and those of the following respiratory variables: VC, FEV_1, and FVC (r = −0.228; FEV_1, r = −0.196; FVC, r = −0.241, respectively), and also between z-scores of abdominal skinfold and VC (r = −0.201). These relations are shown in detail in Table 2.

A stepwise multiple regression analysis was conducted to seek the underlying cause of variations in respiratory function (dependent variables), with body size and shape taken as independent parameters. The VC was best predicted by body height (b = 0.484) and WHR (b = 0.202); the two features explained 32 % of variation in VC. Body height (b = 0.525) and weight (b = −0.188) were significantly associated with FEV_1; the two features explained 26 % of variation in FEV_1. Body height (b = 0.513) and subscapular skinfold (b = −0.163) explained 31 % of variation in FVC. The PEF was predicted by height (b = 0.201), triceps skinfold (b = −0.264), and subscapular skinfold (b = 0.430); with only height and subscapular skinfold significantly associated with PEF. Overall, this set of parameters explained 11 % of variation in PEF. Five parameters of body size and shape explained 20 % of variation in FET, including those significantly associated with FET, such as chest circumferences (b = 2.369), subscapular skinfold (b = −0.551), and others not significantly associated with FET such as waist circumference (b = −1.937), WHR (b = −0.939), and body mass (b = −0.375). The MEF_{75} could not be predicted by BMI (b = −0.188) that explained just 6 % of its variation. There were two parameters with the strongest influence on respiratory function – the body height affecting VC, FEV_1, FVC, and PEF, and the subscapular skinfold affecting FVC, PEF, and FET. The results of stepwise multiple regression analysis are shown in detail in Table 3.

4 Discussion

The aim of the present study was to evaluate the effect of body size and shape and fat distribution in adult women on respiratory function. We found significant correlations between body height and VC, FEV_1, FVC and between the

Table 1 Correlation between age and morphological parameters and respiratory variables

					Circumference			Skinfold		
	Weight	**Height**	**BMI**	**WHR**	**Chest**	**Waist**	**Hip**	**Subscapular**	**Triceps**	**Abdominal**
r	0.24*	−0.50*	0.40*	−0.54*	0.43*	0.50*	0.29*	0.29*	0.17	0.32*
p	0.01	0.001	0.001	0.001	0.001	0.001	0.003	0.002	0.089	0.001
Respiratory function variables										
		VC	**FEV_1**			**FVC**	**PEF**	**MEF_{75}**		**FET**
r		−0.78*	−0.80*			−0.79*	−0.47*	−0.52*		0.16
p		0.001	0.001			0.001	0.001	0.001		0.105

BMI body mass index, *WHR* waist-to-hip ratio, *VC* vital capacity, *FEV1* forced expiratory volume in 1 s, *FVC* forced vital capacity, *PEF* peak expiratory flow, MEF_{75} maximal expiratory flow at 75 % of FVC, *FET* forced expiratory time

*$p < 0.05$

Table 2 Correlation between z-score of morphological and respiratory variables

	VC	FEV_1	FVC	PEF	MEF_{75}	FET
Weight	−0.028	−0.065	−0.002	−0.083	−0.133	0.132
	p = 0.779	p = 0.509	p = 0.982	p = 0.394	p = 0.171	p = 0.174
Height	0.524*	0.481*	0.538*	0.128	0.110	0.023
	p = 0.001	p = 0.001	p = 0.001	p = 0.191	p = 0.259	p = 0.814
BMI	−0.204*	−0.220*	−0.179	−0.119	−0.172	0.125
	p = 0.035	p = 0.023	p = 0.065	p = 0.244	p = 0.077	p = 0.200
WHR	0.250*	0.188	0.170	0.136	0.074	−0.098
	p = 0.009	p = 0.053	p = 0.079	p = 0.162	p = 0.312	p = 0.316
Circumference						
Chest	−0.107	−0.119	−0.050	−0.074	−0.142	0.204*
	p = 0.272	p = 0.222	p = 0.611	p = 0.447	p = 0.146	p = 0.035
Waist	−0.161	−0.149	−0.090	−0.099	−0.140	0.190
	p = 0.097	p = 0.126	p = 0.359	p = 0.312	p = 0.150	p = 0.051
Hip	−0.127	−0.169	−0.120	−0.085	−0.165	0.112
	p = 0.193	p = 0.083	p = 0.217	p = 0.383	p = 0.090	p = 0.250
Skinfold						
Triceps	−0.132	−0.158	−0.131	−0.146	−0.107	0.071
	p = 0.176	p = 0.104	p = 0.178	p = 0.133	p = 0.274	p = 0.467
Subscapular	−0.228*	−0.196*	−0.241*	−0.020	−0.090	−0.005
	p = 0.018	p = 0.043	p = 0.012	p = 0.836	p = 0.358	p = 0.958
Abdominal	−0.201*	−0.170	−0.146	−0.077	−0.088	0.094
	p = 0.038	p = 0.081	p = 0.135	p = 0.434	p = 0.367	p = 0.337

BMI body mass index, *WHR* waist-to-hip ratio, *VC* vital capacity, *FEV1* forced expiratory volume in 1 s, *FVC* forced vital capacity, *PEF* peak expiratory flow, MEF_{75} maximal expiratory flow at 75 % of FVC, *FET* forced expiratory time
*$p < 0.05$

chest circumference and FET. Thus, basic body features strongly influence lung function; so that the higher the body height the better lung function. A stepwise multiple regression analysis confirmed a role of body height in shaping lung function, assessed by spirometry, in women. These results are in line with the effect of body height on FEV_1 and FVC noted in other studies Kozieł et al. (2007).

Body size and shape, and the amount and type of fat distribution are reflected in the occurrence of overweight and obesity. The effect of overweight on respiration can have health consequences, as it can impair lung function and thus increase mortality due to a variety of disorders such as respiratory diseases, cancers, ischaemic heart disease, stroke, and other causes (Hole et al. 1996). Irregularities in lung function have been found in obese children (Li et al. 2003). The awareness of the impact of obesity on health can contribute to changes in lifestyle, as studies show improvement in respiratory activity and a reduction in mortality due to proper diet. Bottai et al. (2002) have shown that weight reduction in the obese contributes to the improvement of lung functioning, and conversely a significant increase in weight decreases lung functions.

We also found that increases in the WHR index correlated with increased VC in women. The WHR ratio reflects the body shape and fat distribution combined. Moreover, WHR with waist circumference is a good indicator of visceral fat, which is often associated with metabolic and cardiovascular diseases (McClean et al. 2008; Harik-Khan et al. 2001). The waist circumference in women is often connected with gynoid fat distribution (Bray 2004). Harik-Khan et al. (2001) have reported a negative interaction between WHR ratio and FEV_1 in men, but not in women. The present findings, in contrast, show the presence of this interaction also in women. It

Table 3 Results of stepwise multiple regression analysis with respiratory functional variables as dependent variables

z-score	Beta	SE of Beta	p
Dependent variable VC ($R^2 = 0.32$; F(3.10) = 16.18; $p < 0.001$)			
Body height	0.484*	0.084	0.001
WHR	0.202*	0.082	0.015
Dependent variable FEV_1 ($R^2 = 0.26$; F(2.10) = 18.68; $p < 0.001$)			
Body height	0.525*	0.087	0.001
Body mass	−0.188*	0.087	0.032
Dependent variable FVC ($R^2 = 0.31$; F(2.10) = 23.97; $p < 0.001$)			
Body height	0.513*	0.082	0.001
Subscapular skinfold	−0.163*	0.082	0.049
Dependent variable PEF ($R^2 = 0.11$; F(6.10) = 2.08; $p < 0.063$)			
Body height	0.201*	0.100	0.048
Waist circumference	−0.316	0.170	0.066
Triceps skinfold	−0.264	0.137	0.057
Subscapular skinfold	0.430*	0.183	0.020
Dependent variable MEF_{75} ($R^2 = 0.06$; F(3.10) = 2.47; $p < 0.066$)			
BMI	−0.188	0.096	0.051
Dependent variable FET ($R^2 = 0.20$; F(7.99) = 3.60; $p < 0.002$)			
Body mass	−0.375	0.239	0.110
WHR	−0.939	0.519	0.073
Chest circumference	2.369*	1.079	0.031
Waist circumference	−1.937	1.353	0.155
Subscapular skinfold	−0.551*	0.167	0.001

VC vital capacity, *FEV_1* forced expiratory volume in 1 s, *FVC* forced vital capacity, *PEF* peak expiratory flow, *MEF_{75}* maximal expiratory flow at 75 % of FVC, *FET* forced expiratory time
*$p < 0.05$

seems that the WHR ratio in women may reflect the overall body stature and that is why it shows a positive correlation with VC. It remains to be explored whether the WHR ratio might be a reliable trait in studies on the relationship between fat distribution and respiratory functions. Waist circumference, which is a component of waist-to-hip ratio did not show a significant correlation with any respiratory variable in the present study. Nonetheless, stepwise multiple regression demonstrated that waist circumference is an element influencing PEF and MEF_{75} flows.

Our findings show significant negative correlations between BMI and VC and FEV_1, between the subscapular skinfold and VC, FEV_1, FVC, and also between abdominal skinfold and VC. Hence, the higher relative contribution of body mass to body height evaluated by BMI and the higher body fat evaluated by subscapular and abdominal skinfold thickness, the worse is the functioning of respiratory system of women. The BMI is an indicator of nutritional status that is strongly associated with the amount of adipose tissue (McClean et al. 2008). The relationship between BMI and respiratory function has not yet been clearly determined. Some studies indicate a negative effect of high BMI on respiration (Lessard et al. 2011; Babb et al. 2008), but others show no such effect (Lazarus et al. 1998). Cotes et al. (2001) have emphasized a limited and ambiguous usefulness of BMI in the assessment of lung function, particularly without considering the amount of adipose tissue and its distribution. Lean individuals may have considerably larger amount of adipose tissue than it results from the relative contribution of body mass to body height (BMI). In contrast, there may be individuals among people of heavyset build who develop musculature and high values of body mass not related to adipose tissue excess (Kozieł et al. 2007), which underscores the necessity lung function in addition to BMI index while examining obesity (Lessard et al. 2011). It is worth noting that the findings of the present study failed to show an unambiguous association between body mass and respiratory variables. Stepwise multiple regression showed, however, that body mass negatively affects FEV_1.

In this study we also found that the subscapular and abdominal skinfolds, indicators of fat distribution, were negatively related with respiratory variables, which shows that upper body fatness, especially abdominal and back fatness, has an impact on respiratory function. Similar results have been reported by Lazarus et al. (1998) and by Collins et al. (1995) who show negative effects of subscapular and upper body fatness on respiratory function, respectively, although the strongest negative influence on respiratory function exerts biceps skinfold thickness. According to Kozieł et al. (2007), a negative influence of upper body fat on FVC and FEV_1 is observed in women; in whom the main contributing factor is subcutaneous chest fatness, as opposed to subcutaneous and visceral abdominal fatness and chest fatness in men. Defining the exact body area whose excessive fatness affects respiratory system is hardly possible considering the ambiguous results above outlined. The present findings failed to substantiate any effect of the triceps skinfold thickness but confirmed a strongest influence on respiratory function of upper body fatness, mainly abdominal and back area.

5 Summary

In this study we indicated the impact of body size and shape characteristics on respiratory variables in women:

1. Body height and chest circumference was positively associated with VC, FEV_1, and FVC, while body mass negatively affected the FEV_1. The values of VC and FEV_1 decreased with increased BMI.
2. Body shape, determined by the WHR indicator, was positively associated with VC, beside that we failed to substantiate any appreciable relation of WHR to respiratory variables.
3. We found a significant negative influence of subscapular fatness on VC, FEV_1, and FVC, and of abdominal fatness, assessed from the skinfold measurements, on VC.

Associations between respiratory variables and morphological features observed in the present study imply a substantial influence of body structure and body fat distribution on lung function. It seems worth considering to take into account body composition, such as the content of adipose tissue and lean body mass, in the routine assessment of lung function by spirometry.

Acknowledgments This work was funded by Ministry of Science and Higher Education of the Republic of Poland, from the quality promoting subsidy, under the Leading National Research Centre (KNOW) programme for the years 2014–2019.

Conflicts of Interest The authors declare no conflicts of interest in relation to this article.

References

Babb TG, Wyrick BL, DeLorey DS, Chase PJ, Feng MY (2008) Fat distribution and end-expiratory lung volume in lean and obese men and women. Chest 134:704–711

Bottai M, Pistelli F, Di Pede F, Carrozzi L, Baldacci S, Matteelli G, Scognamiglio A, Viegi G (2002) Longitudinal changes of body mass index, spirometry and diffusion in a general population. Eur Respir J 3:665–673

Bray GA (2004) Medical consequences of obesity. J Clin Endocrinol Metabol 89:2583–2589

Chen Y, Rennie D, Cormier YF, Dosman J (2007) Waist circumference is associated with pulmonary function in normal-weight, overweight, and obese subjects. Am Soc Clin Nutr 85:35–39

Collins LC, Hoberty PD, Walker JF, Fletcher EC, Peiris AN (1995) The effect of body fat distribution on pulmonary function tests. Chest 107:1298–1302

Cotes JE, Chinn DJ, Reed JW (2001) Body mass, fat percentage, and fat free mass as reference parameters for lung function: effects on terms for age and sex. Thorax 56:839–844

Gauderman WJ, McConnell R, Gilliland F, London S, Thomas D, Avol E, Vora H, Berhne K, Rappaport

EB, Lurmann F, Margolis HG, Peters J (2000) Association between air pollution and lung function growth in southern California children. Am J Respir Crit Care Med 162:1383–1390

Harik-Khan RI, Wise RA, Fleg JL (2001) The effect of gender on the relationship between body fat distribution and lung function. J Clin Epidemiol 54:399–406

Hole DJ, Watt GC, Davey-Smith G, Hart CL, Gillis CR, Hawthorne VM (1996) Impaired lung function and mortality risk in men and women: findings from the Renfrew and Paisley prospective population study. BMJ 313:711–715

Kozieł S, Ulijaszek SJ, Szklarska A, Bielicki T (2007) The effects of fatness and fat distribution on respiratory functions. Ann Hum Biol 34:123–131

Lazarus R, Gore CJ, Booth M, Owen N (1998) Effects of body composition and fat distribution on ventilatory function in adults. Am Soc Clin Nutr 68:35–41

Lessard A, Alméras N, Turcotte H, Tremblay A, Després JP, Boulet LP (2011) Adiposity and pulmonary function: relationship with body fat distribution and systemic inflammation. Clin Invest Med 34(2):E64–E70

Li AM, Chan D, Wong E, Yin J, Nelson EA, Fok TF (2003) The effects of obesity on pulmonary function. Arch Dis Child 88:361–363

Martin R, Saller K (1957) Lehrbuch der Anthropologie, 3rd edn. G. Fischer Verlag, Stuttgart, pp 1957–1959

McClean KM, Kee F, Young IS, Elborn JS (2008) Obesity and the lung: 1. Epidemiology. Thorax 63(7):649–654

Melo LC, Silva MA, Calles AC (2014) Obesity and lung function: a systematic review. Einstein (Sao Paulo) 12 (1):120–125

Saliman JA, Benditt JO, Flum DR, Oelschlager BK, Dellinger EP, Goss CH (2008) Pulmonary function in the morbidly obese. Surg Obes Relat Dis 4(5):632–639

Sandford AJ, Chagani T, Weir TD, Connett JE, Anthonisen NR, Paré PD (2001) Susceptibility genes for rapid decline of lung function in the lung health study. Am J Respir Crit Care Med 163(2):469–473

Suglia SF, Gryparis A, Schwartz J, Wright RJ (2008) Association between traffic-related black carbon exposure and lung function among urban women. Environ Health Perspect 116:1333–1337

Advs Exp. Medicine, Biology - Neuroscience and Respiration (2016) 21: 63–70
DOI 10.1007/5584_2015_207

Published online: 29 January 2016

Concurrent Validity and Reliability of a New Balance Scale Used in Older Adults

Oz Zur, Tamar Shaki, and Eli Carmeli

Abstract

Adults over the age of 70 are at risk of falling. Various balance tests have been developed to identify balance dysfunctions. Their disadvantages including ceiling effects and low sensitivity and duration led to the development of a new balance test. The present study was conducted to determine the concurrent validity, reliability, sensitivity, and specificity of the Zur Balance Scale (ZBS). In this descriptive, cross-sectional study, 76 senior adults were recruited from an independent senior living community and were administered the Berg Balance Scale (BBS) and the ZBS. The BBS was used as the standard of comparison. The ZBS includes head movements and time to maintain to balance. All the subjects completed the tests. Concurrent validity was $r = 0.782$ ($p < 0.0001$). The ZBS had high intra-test (0.897) and inter-test (0.934) correlation coefficients. Its sensitivity was 60 % and specificity 91 % for identifying falls. The dynamic portions of the ZBS capture the integration of the visual, vestibular, and somatosensory systems, as it mimics dynamic spatial aspects of daily activities. We conclude that the ZBS is reliable compared with BBS. It is a simple, easy to administer test that may predict future risk of falls.

Keywords

Balance • Balance testing • Reliability • Validity • Vestibular

O. Zur (✉)
Department of Physical Therapy, Ben Gurion University of the Negev, Beer Sheva, Israel

The Israeli Center for Dizziness and Balance Disorders, 142 Ahuza St, Ra'anana 4330010, Israel
e-mail: zurbalance@gmail.com

T. Shaki
Department of Physical Therapy, Ben Gurion University of the Negev, Beer Sheva, Israel

E. Carmeli
Department of Physical Therapy, University of Haifa, 199 Aba Khoushy Ave, Haifa, Israel

1 Introduction

Falls among older adults have been associated with hospitalizations, institutionalization, fear of falling, greater risk for future falls, increased dependency, decreased mobility, and early mortality (Roe et al. 2009). The annual incidence of falls in adults over the age of 70 is 1 in 3. The ratio rises to 1 in 2 over the age of 85 (de Castro et al. 2015; Boulgarides et al. 2003). The incidence of vestibular dysfunction increases with age and it is 84 % after age 80 (Agrawal et al. 2009).

Falling is a multifactorial phenomenon with intrinsic and extrinsic features. Balance or gait disorders, dizziness/vertigo, confusion, postural hypotension, visual impairment, and unexpected accidents are among the most common causes (de Castro et al. 2015; Rubenstein 2006; Weiss et al. 2013). Clinicians need tests that can help identify those at risk of falling and that may determine the main factors responsible for the balance limitation in order to choose optimal and early interventions (Horak 1997). Many different methods for measuring balance in older adults, with the goal of predicting values for fallers and non-fallers have been developed with moderate-to-high inter-tester reliability, such as the Berg Balance Scale (BBS) (Berg et al. 1989; La Porta et al. 2012; Muir et al. 2008), the Timed Up and Go Test (Boulgarides et al. 2003), and the Functional Reach Test (Lin et al. 2012). The mini BEST test is a new instrument that includes 14 balance tasks to identify various limitations in postural control (King et al. 2012). Its major advantage lies in a comprehensive approach; yet it lacks specific test conditions for head movements during standing, in order to task the vestibular system. It takes 10–15 min to administer. Likewise, the Modified Clinical Test Sensory Interaction for Balance (mCTSIB) is a well-known test that considers the primary functions for balance. It has four test conditions, but does not include head movements (Park et al. 2013). The Dynamic Gait Index (Whitney et al. 2003) addresses the items related to head movements in the pitch and yaw planes, but these assessments are done while walking.

Similar to other investigations, for the purpose of this study the BBS was chosen to serve as the standard of comparison. The main advantages of the BBS are that it is quick to administer (about 14 min), uses easily-acquired equipment, and involves simple functional tasks. Due to its high reliability and validity, the BBS is used to establish concurrent validity and is often used in research to assess treatment outcomes and as a validation instrument for other balance assessment tools (Langley and Mackintosh 2007; O'Sullivan et al. 2009; Geiger et al. 2001). The BBS is reproducible and has good inter-tester reliability (La Porta et al. 2012). The test is more appropriate for participants with moderate-to-severe balance dysfunction. Yet a drawback of the BBS is that it has a fairly low sensitivity to detect change in a patient's balance over time. It also has a ceiling effect (Stevenson 2001), although less so than other balance scales, such as the Performance-Oriented Mobility Assessment Tool or the Dynamic Gait Index (Pardasaney et al. 2012; Whitney et al. 2003).

The Zur Balance Scale (ZBS) is a new tool designed to evaluate balance. It measures the effects of the three main sensory systems (visual, vestibular, and somatosensory) operating together to maintain balance. Horizontal and vertical head movements are used specifically to assess the dynamic aspects of the vestibular system. The ZBS measures balance while the participant is standing on a firm surface or a half cylinder of styrofoam, in the tandem or Romberg position. It takes only 4–5 min to administer and uses simple, easily-acquired equipment.

The purpose of this study was to determine the concurrent validity, reliability, and sensitivity and specificity (i.e., to predict future falls) of the ZBS by comparing it with the BBS. The ZBS includes head movements while standing in varied positions on different surfaces.

2 Methods

2.1 Zur Balance Scale (ZBS)

The present study was approved by the Institutional Review Board of Maccabi Health Maintenance Organization (permit no. 14/2014). All participants provided written, informed consent.

The ZBS is a screening test for assessing balance function. It is quick (4–5 min) and simple to administer and analyze. Equipment needed for the test is a half-cylinder of styrofoam 60 cm long × 18 cm wide × 9 cm high, a stop watch for measuring time in seconds, and a metronome set at one Hz. The styrofoam has a density of 30 kg/m^3 and is covered tightly with a stretchable piece of fabric. The ZBS should be conducted in a quiet room. The tested participant is asked to stand 2 m from the fixed target, a 5 × 5 cm X mark at the eye level (±30°). A solid support (such as a chair or table) is placed next to the participant for safety and confidence, while the examiner stands in front of the participant, to the side. Participants are asked to stand consecutively in Romberg or tandem stance on the floor or on the styrofoam while completing a series of four different tasks (eyes open, eyes closed, horizontal head movements, and vertical head movements).

Each combination of stance and task comprises a different condition, for a total of ten conditions evaluated. The ability to maintain balance for a maximum of 10 s is measured for each condition. The test begins with the participant standing stable on the floor. This can be achieved with a support on one side and the examiner on the other. The test is started with participant's hands on his/her hips, when the participant is ready. Each condition is performed twice and the better of the two is recorded for analysis.

The ZBS is scored by counting the number of head movements (HM) and time to maintain balance. The time to maintain balance (with and without HM) is measured in seconds, for a maximum of 10 s (Table 1, black boxes indicate time without head movements). In 5 of the 10 conditions (2, 3, 6, 8, 10; white boxes), the participant is asked to move his/her head left and right covering an arc of approximately 120° (60° to each side) and a total of 60° up and down (30° up and 30° down), each within 10 s according to a 60 Hz metronome. From 0 to 10 HM are performed in each condition, for a maximum of 50. The ZBS score is calculated by summing the total number of HM multiplied by 2, plus the total time (in seconds) divided by 2.

Table 1 Zur Balance Scale – score sheet

Condition	Task	Abbreviation	Head motion	Time (s)
1	Romberg stance on the floor, eyes closed	(ROM_EC)		
2	Romberg stance on the floor, during horizontal head movements, eyes closed	(ROM_HM)		
3	Romberg stance on the floor, vertical head movements, eyes closed	(ROM_VM)		
4	Tandem stance on the floor, with eyes open on a fixed target	(TAN_EO)		
5	Tandem stance on the floor, eyes closed	(TAN_EC)		
6	Tandem stance on the floor, during horizontal head movements	(TAN_HM)		
7	Romberg stance on styrofoam, eyes open on a fixed target	(S_ROM_EO)		
8	Romberg stance on styrofoam with vertical head movements, eyes open	(S_ROM_VM)		
9	Tandem stance on styrofoam, with eyes open on a fixed target	(S_TAN_EO)		
10	Tandem stance on styrofoam, with horizontal head movements, eyes open	(S_TAN_HM)		
Total	Score Calculation		(Head movements× 2) Max = 100	Sum of sec Max = 100

ROM Romberg stance, *EC* eyes closed, *EO* eyes open, *HM* horizontal head movements, *VM* vertical head movements, *TAN* Tandem stance, *S_ROM* Romberg stance on Styrofoam, *S_TAN* Tandem stance on styrofoam

The BBS was conducted according to the protocol described by Berg et al. (1989).

2.2 Study Protocol

In this descriptive, cross-sectional, double blind study, 300 older adults residing in an independent living community were invited to participate in a lecture entitled 'Balance and Falls'. They were introduced to the ZBS and to the BBS. Inclusion criteria were age 70 years or over and ability to walk independently, with or without a cane. Following the lecture, 110 volunteered to participate in the study and signed a consent form. A total of 76 subjects of the mean age of 83 ± 5 years, range 71–97 years, met the inclusion criteria. Sixty of them (79 %) were female. The participants lived in the independent senior living community for a mean of 3 ± 1.5 years. They had an average of 12 ± 3 years of education and were engaged in sport activities for a median of 3 h a week.

Sociodemographic data were collected including date of birth, gender, fall history, fall-related injury, physical exercise activity, social activity, and the length of residence in the facility. Exclusion criteria included assistive device for standing, a static visual deficit (i.e., unable to read at least the first five lines on the Snellen eye chart even with vision correction), cognitive deficit (Mini-Mental State Examination score of less than 24), neurological condition (such as Parkinson's disease or cerebrovascular accident), or acute orthopedic conditions (such as hip fracture).

Participants were randomly administered the ZBS and the BBS on the same day (T1) by two experienced clinical physical therapists. One physical therapist administered the ZBS (tester 1) and another administered the BBS (tester 2) to evaluate the validity of the ZBS. For reliability testing, the ZBS was readministered by the same physical therapist, under the same conditions (i.e., time of day and place) 10 days later (T2). In addition, to evaluate inter-tester reliability, the ZBS was also administered by a third, experienced clinical physical therapist (first author). Thus, each participant was tested twice at T2 and the order of the therapists also was randomized.

The medical staff of the independent living community maintains strict fall monitoring and surveillance policies. Falls during the 18 months after the balance examinations were collected from the medical records as documented in a report by the faller or by a significant other, usually the medical staff. This follow-up information was used to determine the cut-off point for the likelihood of falling (see the section on sensitivity and specificity for fall prediction below).

2.3 Statistical Analysis

Two previous studies that compared new balance tests to the BBS were used to determine the minimum sample size (Langley and Mackintosh 2007; Whitney et al. 2003). Based on the numbers reported in those studies, we planned to enroll a minimum of 70 participants (Roe et al. 2009).

The BBS score was converted into a percentage and the ZBS was measured on a numerical scale from 0 to 100 in order to have comparative scales. The BBS was used as the standard for establishing concurrent validity. Concurrent validity of the ZBS was assessed against the BBS with Pearson's correlation of the ZBS against the BBS. Test-retest in two different sessions and inter-tester reliability were assessed using intra-class correlations ICC.

Receiver operating characteristic (ROC) was used to determine the cut-off scores of the BBS (not presented) and the ZBS between fallers and non-fallers. Specificity and sensitivity were calculated. Differences between nominal parameters and fall status were calculated using the Chi-squared test. Differences between continuous variables were calculated using a *t*-test. $P < 0.05$ was considered statistically significant. All analyses were performed using IBM, SPSS-22 software.

3 Results

During the follow-up of testing, 13 participants (17 %) experienced a fall (eight had one fall and five had at least two falls). There were no statistical differences in the background parameters of age, gender, years of education, years of residence in the facility for seniors, or exercise activity between fallers and non-fallers.

3.1 Concurrent Validity, Intra-tester Reliability, and Inter-tester Reliability

The mean ZBS score was 55 ± 12.8 (min 6, max 82, median 56). The mean BBS score was 87 ± 13.2 (min 7, max 100, median 91). Validity was indicated by Pearson's correlation between the ZBS and the BBS ($r = 0.682$, $p < 0.0001$) (Fig. 1).

The ZBS was administered twice by the same tester at 10–14 day intervals to evaluate intra-session reliability. In addition, the ZBS was randomly administered by a third tester to evaluate inter-tester reliability. The intra-tester reliability ICC was 0.934 (95 % CI = 0.904–0.956) and inter-tester reliability ICC was 0.934 (95 % CI = 0.904–0.956).

3.2 Sensitivity and Specificity for Fall Prediction

The ROC curve was used to find the cut-off score of the ZBS to predict falls (Fig. 2). The cut-off point was 0.56 as an optimal point to predict falls. The area under the curve was 0.755, (95 % CI = 0.615–0.895). For comparison, the cut-off point of the BBS was 0.90. Both BBS and ZBS were used to predict an individual's faller status. The ZBS's sensitivity was 60 % and specificity was 91 %. The BBS's sensitivity was 66 % and specificity was 91 %.

4 Discussion

The purpose of this cross-sectional study was to determine the concurrent validity, reliability, and sensitivity and specificity of the Zur Balance Scale (ZBS), a new scale to evaluate dynamic vestibular function among older adults. The ZBS was developed over several years based on a content validity assessment process among 30 experienced physical therapists, researchers, and neuroethology physicians and 20 years of experience working with thousands of individuals with dizziness and balance disorders. A half-cylinder of styrofoam was used for 4 of

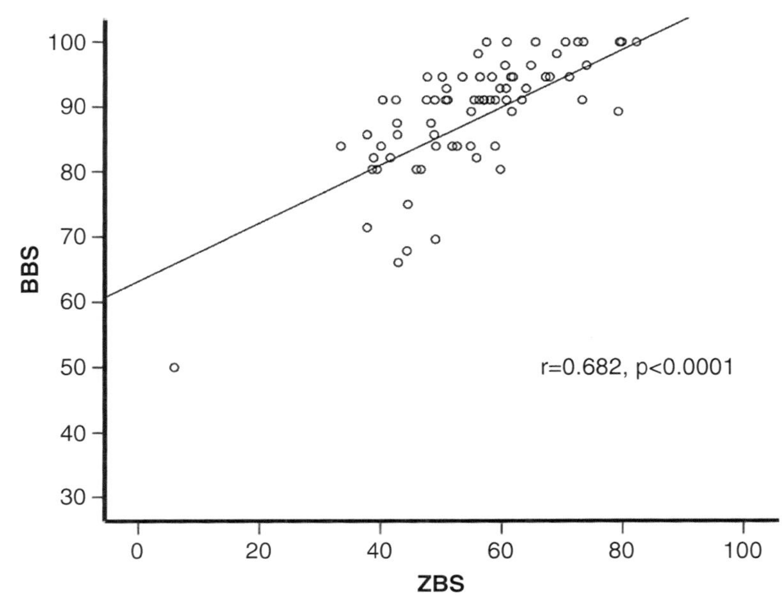

Fig. 1 Concurrent validity between Berg Balance Scale (BBS) and Zur Balance Scale (ZBS)

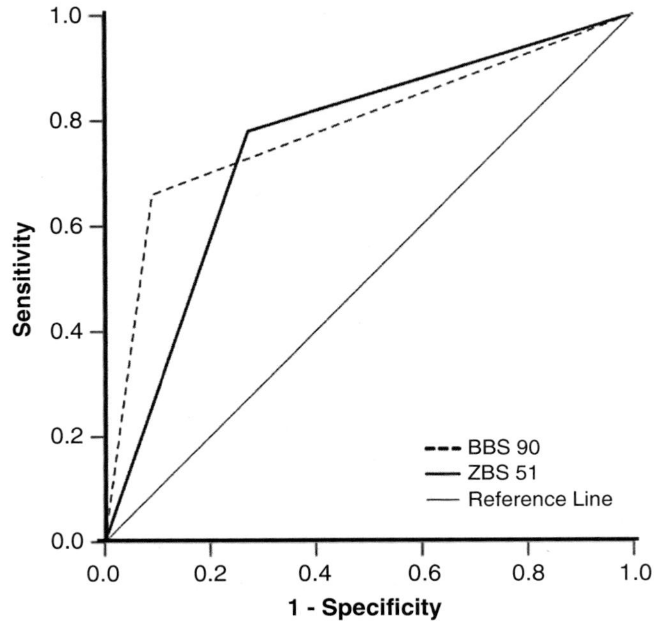

Fig. 2 Receiver operating curves (ROC) for Berg Balance Scale (BBS) and Zur Balance Scale (ZBS)

the 10 items on the ZBS to alter somatosensory input. The cylinder's length and width were chosen to suit the foot size of most people and the height to maintain safety.

To assess the concurrent validity of the ZBS, the scale had to be compared with a well-known, validated, and reliable test, such as the BBS. The BBS is usually the first choice for assessing balance among older adults (Berg et al. 1989; La Porta et al. 2012). We chose not use the Mini BESTest, even though it includes horizontal and vertical head movements, but does so only while walking. We did not use the mCTSIB either, since it does not include head movements at all.

The results demonstrate that in some aspects the ZBS is as good as the BBS for evaluating older adults. As a new balance test, the ZBS is important because it includes horizontal and vertical head movements during different stances, whereas other balance tests do not consider head motions while standing still. The Fullerton Advanced Balance Scale (Rose et al. 2006) and the Dynamic Gait Index (Whitney et al. 2003) request participants to perform head movements during gait examination. The assessment of a static balance with head movements is indispensable, since this evaluates the sensory systems involved in maintaining balance (Malstrom et al. 2007).

The ZBS was found to be as reliable as the BBS in intra-session reliability and inter-tester reliability. Both tests are both important tools for therapists assessing balance. However, the added value of ZBS is that it focuses on the dynamic function of the vestibular system. Thus, vestibular impairments may be more easily identified by using ZBS.

A few limitations to the study should be noted. We did not perform logistic regressions with other interacting variables, such as the level of physical activity, medications used, and comorbidities. One possible weakness of this study might be using the styrofoam for 4 of the 10 items. Styrofoam might have a potential to reduce foot contact; thereby introducing a confounding variable, i.e., increased requirement for a hip strategy to control one's center of mass over the base of support. Therefore, future studies may also compare ZBS with mCTSIB on a force plate. We also found a lower proportion of fallers (17 %) than would be expected. Despite the fact that a fall monitoring and surveillance program was rigorously managed, this low frequency of fall events might suggest that mainly falls with injuries were recorded and some minor falls might have been missed.

We believe that older adults with a feeling of imbalance should be evaluated by ZBS, and not by BBS which does not include measures that stress vestibular function. Since vestibular dysfunction is common among older adults, the ZBS should be administered first. On the other hand, BBS should be the first choice for assessing older adults with severe balance deficits, since many of the test conditions are easier to perform compared to ZBS. In our opinion, ZBS should be the first choice for independent older adults. After 18 months of follow-up of falls in the study population, the ZBS seems to be a sensitive test for detecting balance dysfunction and predicting falls.

In conclusion, ZBS is potentially equivalent to BBS for balance assessment. The ZBS highlights the integration of the three main sensory systems involved in maintaining balance. Specifically, it mimics the dynamic, spatial aspects of daily activities such as standing on an uneven surface with voluntary head movements. In addition, ZBS is quick to administer and the score is easy to calculate on a 0–100-point scale. The ZBS can be used to assess participants before, during and after vestibular rehabilitation.

Acknowledgements The author would like to thank Profs. Cortney Hall and Michael Shubert for their comments on the previous files. Special thanks go to the physical therapists Yakir Ariam and Yair Ohel who performed the examinations and to the professional staff of Beit Ildan. The manuscript was edited by Faye Schreiber. Nava Jelin performed the statistical analysis.

Conflicts of Interest This study was not supported by external funding. The authors have no conflicts of interest to declare.

References

Agrawal Y, Carey JP, Della Santina CC, Schubert MC, Minor LB (2009) Disorders of balance and vestibular function in US adults: data from the National Health and Nutrition Examination Survey, 2001–2004. Arch Intern Med 169(10):938–944

Berg K, Wood-Dauphine S, Williams JI, Gayton D (1989) Measuring balance in the elderly: preliminary development of an instrument. Physiother Can 41(6):304–311

Boulgarides LK, McGinty SM, Willett JA, Barnes CW (2003) Use of clinical and impairment-based tests to predict falls by community-dwelling older adults. Phys Ther 83(4):328–339

de Castro V, Mokoroa O, Artieda J, Muniozguren N, Etxebarriarteun L, Alvarez L, Garcia Calabuig MA, Red de MédicosVigía del País Vasco (2015) Epidemiology of accidents in a cohort of adults over 64 years old in the Autonomous Community of the Basque Country. Rev Esp Geriatr Gerontol 50(6):281–284 (Article in Spanish)

Geiger AG, Allen JB, O'Keefe J, Hicks RR (2001) Balance and mobility following stroke: effects of physical therapy interventions with and without biofeedback/forceplate training. PhysTher 81(4):995–1005

Horak FB (1997) Clinical assessment of balance disorders. Gait Posture 6(1):76–84

King LA, Priest KC, Slarin A, Pierce D, Horak FB (2012) Comparing the Mini-BESTest with the Berg Balance Scale to evaluate balance disorders in Parkinson's disease. Park Dis 2012:375–419

La Porta F, Caselli S, Susassi S, Cavallini P, Tennant A, Franceschini M (2012) Is the Berg Balance Scale an internally valid and reliable measure of balance across different etiologies in neurorehabilitation? A revisited Rasch analysis study. Arch Phys Med Rehabil 93 (7):1209–1216

Langley FA, MacKintosh SFH (2007) Functional balance assessment of older community dwelling adults: a systematic review of the literature. Internet J Allied Health Sci Pract 5(4):1–11

Lin YH, Chen TR, Tang YW, Wang CY (2012) A reliability study for standing functional reach test using modified and traditional rulers. Percept Mot Skills 115 (2):512–520

Malstrom EM, Karlberg M, Melander A, Magnusson M, Moritz U (2007) Cervicogenic dizziness-musculoskeletal findings before and after treatment and long-term outcome. Disabil Rehabil 29 (15):1193–1205

Muir SW, Berg K, Chesworth B, Speechley M (2008) Use of the Berg Balance Scale for predicting multiple falls in community-dwelling elderly people: a prospective study. Phys Ther 88(4):449–459

O'Sullivan M, Blake C, Cunningham C, Boyle G, Finucane C (2009) Correlation of accelerometry with clinical balance tests in older fallers and non-fallers. Age Ageing 38(3):308–313

Pardasaney PK, Latham NK, Jette AM, Wagenaar RC, Ni P, Slavin MD, Bean JF (2012) Sensitivity to change and responsiveness of four balance measures for community-dwelling older adults. Phys Ther 92 (3):388–397

Park MK, Kim KM, Jung J, Lee N, Hwang SJ, Chae SW (2013) Evaluation of uncompensated unilateral vestibulopathy using the modified clinical test for sensory interaction and balance. Otol Neurotol 34 (2):292–296

Roe B, Howell F, Riniotis K, Beech R, Crome P, Ong BN (2009) Older people and falls: health status, quality of life, lifestyle, care networks, prevention and views on service use following a recent fall. J Clin Nurs 18 (16):2261–2272

Rose DJ, Lucchese N, Wiersma LD (2006) Development of a multidimensional balance scale for use with functionally independent older adults. Arch Phys Med Rehabil 87(11):1478–1485

Rubenstein LZ (2006) Clinical risk assessment, interventions and services: falls in older people: epidemiology, risk factors and strategies for prevention. Age Ageing 35(2)-S2, ii37–ii41

Stevenson TJ (2001) Detecting change in participants with stroke using the Berg Balance Scale. Aust J Physiother 47(1):29–38

Weiss A, Brozgol M, Dorfman M, Herman T, Shema S, Giladi N, Hausdorff JM (2013) Using 3-day accelerometer recordings. Does the evaluation of gait quality during daily life provide insight into fall risk? A novel approach. Neurorehab Neural Repair 27 (8):742–752

Whitney S, Wrisley D, Furman J (2003) Concurrent validity of the Berg Balance Scale and the dynamic gait index in people with vestibular dysfunction. Physiother Res Int 8(4):178–186

Advs Exp. Medicine, Biology - Neuroscience and Respiration (2016) 21: 71–85
DOI 10.1007/5584_2015_203

Published online: 29 January 2016

Quality of Care for Patients with Chronic Respiratory Diseases: Data for Accreditation Plan in Primary Healthcare

Donata Kurpas, Katarzyna Szwamel, and Bożena Mroczek

Abstract

There are scarce reports in the literature on factors affecting the assessment of the quality of care for patients with chronic respiratory diseases. Such information is relevant in the accreditation process on implementing the healthcare. The study group consisted of 133 adult patients with chronic respiratory diseases and 125 adult patients with chronic non-respiratory diseases. In the present study, the level of satisfaction from healthcare provided by the primary healthcare unit, disease acceptance, quality of life, health behaviors, and met needs were examined, as well as associations between variables with the use of correspondence analysis. The results are that in patients with chronic respiratory diseases an increase in satisfaction depends on the improvement of well-being in the mental sphere. The lack of problems with obtaining a referral to a specialist and a higher level of fulfilled needs also have a positive effect. Additionally, low levels of satisfaction should be expected in those patients with chronic respiratory diseases who wait for an appointment in front of the office for a long time, report problems with obtaining a referral to additional tests, present a low level of health behaviors, and have a low index of benefits.

D. Kurpas (✉)
Department of Family Medicine, Wroclaw Medical University, 1 Syrokomli St., 51-141 Wroclaw, Poland

Opole Medical School, 68 Katowicka St., 45-060 Opole, Poland
e-mail: dkurpas@hotmail.com

K. Szwamel
Independent Public Healthcare Center, Hospital Emergency Ward and Admissions, 2 Roosevelta St., 47-200 Kędzierzyn- Koźle, Poland

B. Mroczek
Department of Humanities in Medicine, Faculty of Health Sciences, 11 Gen. Dezyderego Chłapowskiego St., 70-103 Szczecin, Poland

Keywords
Care program • Chronic disease • Healthcare delivery • Patient satisfaction • Primary care • Pulmonary disease • Quality of healthcare

1 Introduction

The common challenge for healthcare systems in the EU is the issue of growing consequences of chronic diseases of an aging population. It is estimated that the cost of healthcare for chronically ill individuals now accounts for 70–80 % of healthcare costs in the EU. These costs will continue to grow and the number of people aged over 65 will double between 1990 and 2050 (McKee et al. 2015). A framework for the concerted effort for the care for people living with multimorbidity is needed to give these patients an experience of a healthcare system that cooperates to provide the best possible care (e.g., based on the strategy the Chronic Care Model). Patients with multimorbidity often experience inadequate continuity of care and they face problems in accessing the health professionals they trust (Smidth et al. 2013).

Patients' satisfaction is now regarded as an important indicator of healthcare quality, alongside the quality of life, mortality, and health costs, and is of growing interest to health professionals and policymakers (Sebo et al. 2015). Care managers indicated that components of patients' satisfaction: ensuring primary care provider's follow-up, coordinating appropriate services, providing patient's education, ensuring accurate medication, teaching patients the symptoms of acute exacerbations, and building effective doctor-patient relationships have the greatest impact on patients' clinical outcomes and well-being (Carayon et al. 2015). Other patients' satisfaction factors, such as good communication skills, including taking the time to listen and providing clear explanations, are among the qualities that patients most desire. Effective doctor–patient communication predicts better health outcomes, such as symptom resolution, pain control, and physical functioning (Quigley et al. 2013). Incorrect communication between doctors and patients, difficult access to healthcare and treatment, and unclear information regarding the dosage of medicines can decrease the degree of following doctors' recommendations by patients (Kardas et al. 2013). Increased patients' satisfaction with their regular source of primary care, ensuring continuity of care at the primary level and adherence to a primary care provider can reduce the use of costly care because patients who are adherent have fewer visits at emergency departments and fewer hospitalizations, compared with less satisfied ones (Pourat et al. 2015). Patients want healthcare professionals to include them in the decisions about their own health, and the management of their own lives and diseases (Smidth et al. 2013). Patients' perceptions and expectations of a good General Practitioner (GP) can vary widely across cultures, because of the differences in healthcare services in different countries. However, GPs are always expected to be responsive to the patients' expectations and needs (Sultan et al. 2012). These components are an integral part of modern evaluations of patients' satisfaction.

Thus, measuring satisfaction with healthcare services provides the information on meeting the patients' needs and expectations and on the possible sources of discontent. The research findings allow adjusting the system of care for beneficiaries' expectations and indicates the direction of quality enhancement. There are scarce reports in the literature on factors affecting the assessment of the quality of care for patients with chronic respiratory diseases. Such information is relevant in the accreditation process on implementing the healthcare for this group of patients. Therefore, the purpose of this study was to determine the factors shaping the assessment of the quality of visits within the primary healthcare by patients with chronic respiratory diseases and the differences between patients with respiratory and non-respiratory diseases.

2 Methods

The research was conducted in accordance with the principles of the Declaration of Helsinki. The study was approved by the Bioethics Committee of the Medical University in Wroclaw, Poland (approval no. KB-422/2014). The main inclusion criteria were the following: age (at least 18 year) and the diagnosis of at least one respiratory chronic disease.

The study group consisted of 133 adult patients with chronic respiratory diseases. The median age was 64 years (min-max: 20–90) years. Participants for the study were recruited from patients of 90 general practitioners during the period of July 2014–April 2015. The patients who agreed to participate anonymously in the project signed an informed consent form. The patients were given a questionnaire to complete at home and return in a stamped envelope.

Apart from the study group of patients with chronic diseases of the respiratory system – J1 ($n = 133$), there was a control group consisting of patients with other than respiratory chronic diseases – J0 ($n = 125$). The control group was selected from a sample of 1150 patients with a similar sociodemographic structure to that of the study group. The structure was defined by five bicategorical variables due to a small size of the groups: sex (male, female), age (below, equal, or above the median), place of residence (city, village), education (less than high school, high school, or higher education), and marital status (in relationship, single). The compliance of the structures of both groups was verified positively using the Chi-squared test (Chi2 = 0.04, $df = 31$, $p = 1$).

The Patient Satisfaction Questionnaire was used, which was developed on the EUROPEP questionnaire. It consists of the following modules: feelings of the patient during the interview and physical examination performed by a physician; information received by the patient from the doctor about the disease, diagnosis and treatment, health promotion and disease prevention; the level of information available to the physician on the patient's disease history (previous visits); providing information to the patient about the disease management (including details on the necessary control visits, specialist consultations, and hospitalization) with an assessment of the degree of patient's involvement in the process of decision making in further proceedings; evaluation of the level of communication; concern about the social situation of the patient; attention paid to the patient's emotional difficulties; overall assessment of the family physician, and the evaluation of the patient's contact with non-physician employees of the primary healthcare unit. The patient was able to answer 'yes', 'no', or 'sometimes'. The answers were scored in the following way: 'yes' = 2 points, 'sometimes' = 1 point, and 'no' = 0 points. The level of patient's satisfaction with the services provided by the primary healthcare unit was assessed by summing up the score. The possible score range was 0–72 points. Internal consistency of the Patient Satisfaction Questionnaire measured with the α-Cronbach coefficient was 0.94.

The information obtained from patients helped to determine the index of healthcare services, the somatic index, and the hospitalization rate. The index of healthcare services was determined based on the number of identified benefits received during a visit to a doctor. The somatic index was calculated by summing up the values assigned to the somatic symptoms and dividing by 49 (the maximum attainable number of points). Depending on the frequency of somatic symptoms listed by the patient values from 1 (occurs once a year) to 7 (occurs continuously) were given. For each patient, the number of hospitalizations/number of departments was determined for each year. Then, an average of these three indicators was estimated and this value was adopted as an indicator of hospitalization (3 years) for the patient.

A level of met/unmet needs was estimated with the Camberwell Assessment of Need Short Appraisal Schedule. On the basis of 24 questions which characterize 22 needs, the number of needs was set for the patient to answer whether the need was met (1) or not (0); the numerical score was converted to N - needs satisfied or M – needs unsatisfied. The Camberwell index was calculated according to the formula: M/N. Internal consistency of the Modified Short Camberwell Needs Assessment was 0.96 (-α-Cronbach coefficient).

Quality of life was assessed with the Polish version of the World Health Organization Quality of Life Instrument Short Form (WHOQOL-BREF) within four domains: D1-Physical, D2-Psychological, D3-Social relationships, and D4-Environmental. The reliability of the Polish version of the WHOQOL-BREF questionnaire, measured with the α-Cronbach coefficient refers both to the parts evaluating particular domains (results from 0.81 to 0.69) and the questionnaire as a whole (0.90) (Jaracz et al. 2006).

The patients' adaptation to a life with a disease was assessed using the Acceptance of Illness Scale (AIS) adapted to conditions in Poland. The AIS consists of eight statements about negative consequences of the state of health, where every statement is rated on a five-point Likert-type scale (1 denotes poor adaptation to a disease and 5 its full acceptance). The score for illness acceptance is a sum of all points and can range from 8 to 40. Low scores (0–29) indicate the lack of acceptance and adaptation to a disease and the strong feeling of mental discomfort. High scores (35–40), on the other hand, indicate the acceptance of illness, manifested as the lack of negative emotions associated with a disease. The scale can be used to assess the degree of acceptance of every disease. The α-Cronbach coefficient of the Polish version is 0.85 and of the original version is 0.82.

The authors also used the Health Behavior Inventory (HBI) developed by Juczyński (2001). The instrument consists of 24 statements that measure four categories of pro-health behavior: healthy eating habits, preventive behavior, positive mental attitudes, and health practices. The patient marks the frequency of health behaviors on a scale between 1 and 5 (1 – almost never, 5 – almost always). The sum of results from all four scales indicates the score for the general health behavior (range 24–120); the higher the score, the more healthy the behavior. The intensity of a health behavior in a particular category is the sum of all answers in this category divided by six. The HBI internal consistency measured with Cronbach's alpha is 0.85.

2.1 Statistical Elaboration

Shapiro-Wilk's test was employed to verify normality of data distributions. Normal distribution was confirmed for the following variables in group J1: acceptance of illness, QoL in Physical, Psychological and Environmental domains, total pro-health behaviors, and pro-health behavior in the categories of healthy eating habits and health practices. The normally distributed variables in group J0 were as follows: QoL in Physical and Psychological domains, total pro-health behaviors, and pro-health behavior in the categories of healthy eating habits and health practices.

Arithmetic means, standard deviations, medians, and the range of variability (extremes) were calculated for measurable (quantitative) variables, while the frequency (percentage) was determined for qualitative variables. The analysis of qualitative variables was based on contingency tables and the chi-squared or Fisher's exact test. Wilcoxon's rank sum test was carried out for comparisons between two samples, with the null hypothesis that the distribution of x and y differs by a location shift of 0. Spearman's rank correlation test was used to verify correlations between variables. A difference between correlation coefficients was assessed with an unpaired test.

The correspondence analysis was also used, in which categories (points) belonging to clusters are interpreted as related to each other. For the analysis, variables were selected which in one or both study groups correlated significantly with the level of satisfaction. Each variable was first converted to a bicategorial feature to ensure the best subsequent interpretation of clusters on a two-dimensional graph. The analysis was conducted separately for either group, taking the same selected variables into account. The significance of differences was assumed at

Table 1 Sociodemographic data of chronically ill patients (group J1 i J0) and their diagnoses according to ICD-10, and comorbidities

	J1 (n = 133)		J0 (n = 125)	
Age (years)				
Q.25 %	51.8		50.0	
Q.50 %	64.0		64.0	
Q.75 %	73.0		73.0	
Min-max	20–90		19–90	
Gender	n	%	n	%
Women	71	53.4	67	53.6
Men	62	46.6	58	46.4
Place of residence	n[a]	%	n	%
Village	46	34.8	43	34.4
Below 5,000[b]	17	12.9	24	19.2
5,000–10,000[b]	8	6.1	6	4.8
10,000–50,000[b]	40	30.3	24	19.2
50,000–100,000[b]	8	6.1	10	8.0
100,000–200,000[b]	6	4.5	9	7.2
Over 200,000[b]	7	5.3	9	7.2
Education	n[a]	%	n	%
Primary	26	20.0	21	16.8
Vocational	36	27.7	39	31.2
Secondary	34	26.2	33	26.4
Post-secondary	17	13.0	18	14.4
Higher	17	13.0	14	11.2
Marital status	n[a]	%	n	%
Single	16	12.2	18	14.4
Married	79	60.3	75	60.0
Divorced or separated	6	4.6	9	7.2
Widowed	30	22.9	23	18.4
Diagnosis	n[a]	%	n	%
J45 Bronchial asthma	64	48.1	0	0
J43 Pulmonary emphysema	27	20.3	0	0
J44 Other chronic obstructive pulmonary diseases	19	14.3	0	0
J42 Unspecified chronic bronchitis	19	14.3	0	0
J41 Chronic simple and mucous-purulent bronchitis	16	12.0	0	0
J47 Bronchiectasis	7	5.3	0	0
Most common coexisting diseases[c]	n[a]	%	n	%
M47 Spondylosis	39	29.3	22	17.6
I10 Primary hypertension	30	22.6	50	40.0
I70 Atherosclerosis	26	19.5	20	16.0
M15 Osteoarthritis of multiple joints	15	11.3	13	10.4
E11 Non-insulin-dependent diabetes mellitus	13	9.8	17	13.6

Q quartile

[a]numbers in column *n* do not sum up to 133 due to missing data in the questionnaires completed by patients

[b]city/town population

[c]some patients were diagnosed as having at least two pathological entities or chronic diseases which were not within the top 5 most common diseases

$p < 0.05$. The R 3.1.3 for Mac OS X 10.10.4 statistical software was used for all analyses.

3 Results

The sociodemographic data are presented in Table 1, along with the diagnoses of chronic respiratory diseases and the most common co-existing diseases. The median number of chronic diseases among the patients was 3 (min-max: 1–15). The average BMI (body mass index) was 28.9 ± 4.9 kg/m^2 and the median somatic index: 0.4 (min-max: 0.0–1.0). The median index of healthcare services in the study group was 4.7 (min-max: 1.0–52.7) and the median of hospitalizations was 1.0 (min-max: 0.0–14.0). The results of the satisfaction from healthcare provided by the primary healthcare unit in both groups (J1 and J0) are shown in Table 2 and detailed answers for the questions in group J1 – in Table 3.

The majority of patients negatively assessed the following aspects: information provided by the GP about the side effects of medicines taken, help from the GP in coping with health-related fears and concerns, interest of the GP in the patient's material and personal situation (specifics of the source of social support) and that of other members of the patient's family (their health, contact among family members), and the information about prophylactic examinations (e.g., detection of heart diseases, cancer, etc.). The improvement would require the inclusion of patients into a decision-making on the type of treatment and further proceedings, thorough explanation of test results, providing information on further disease course and treatment, principles of a healthy lifestyle, and on the seriousness of a health problem as well as increased interest of the GP in the patient‘s concerns about disease, forbearance toward child patients, taking into account the parents' opinion on the illness of their child, possibility of talking to the doctor and obtaining clarification over the phone, and to show that the patient's opinion is important for the GP. The remaining components of the quality assessment of the primary health center were evaluated highly by more than 7 out of the 10 patients (Table 3).

The results of total satisfaction from healthcare provided by the primary healthcare unit, of the evaluation of disease acceptance, quality of life, and health behaviors, and of the level of met needs in group J1 are shown in Table 4.

3.1 Significant Associations

Low levels of satisfaction were observed in patients in whom no improvement of well-being in the mental sphere was observed within the last 12 months ($r = 0.18$, $p = 0.046$), who were waiting for an appointment in front of the doctor's office for a long time ($r = -0.19$, $p = 0.033$) and according to them it was too long ($r = -0.23$, $p = 0.009$), reporting a

Table 2 Satisfaction with healthcare provided by the primary healthcare unit in both groups (J1 and J0)

Level of satisfaction		J1	J0	Test[a]
low (0–35 points)	n	12	16	
	%	9.0	12.8	$p = 0.599$[b]
medium (36–53 points)	n	30	29	
	%	22.6	23.2	
high (54–72 points)	n	91	89	
	%	68.4	64.0	
Total	n	133	125	
	%	100.0	100.0	

[a]Wilcoxon‘s rank sum test conducted to verify the hipothesis regarding the compliance of distribution in both groups for the original Satisfaction variable (i.e., uncategorized as in this table): $W = 9077.5$, $p = 0.201$. [b]Fisher's Exact test for count data. Both tests did not reject the hipothesis regarding the compliance of distribution between groups J1 and J0

Table 3 Distribution of answers to questions within the Patient Satisfaction Questionnaire (group J1 only, n = 133); n (%)

Variables	No	Sometimes	Yes
GP remembers the patient's earlier visit(s)	8 (6.0)	15 (11.3)	110 (82.7)
GP devotes sufficient time during the visit	5 (3.8)	18 (13.5)	110 (82.7)
GP listens attentively	6 (4.5)	16 (12.0)	111 (83.5)
GP talks to the patient as long as he/she expects	8 (6.0)	28 (21.1)	97 (72.9)
GP carries out the examination carefully	2 (1.5)	19 (14.3)	112 (84.2)
GP is gentle during the examination	1 (0.8)	10 (7.5)	122 (91.7)
GP respects patient privacy	1 (0.8)	9 (6.8)	123 (92.5)
GP gives the name of the disease	6 (4.5)	18 (13.5)	109 (82.0)
GP explains the cause of the symptoms	11 (8.3)	22 (16.5)	100 (75.2)
GP allows the patient to decide on the type of treatment and further proceedings	15 (11.3)	39 (19.3)	79 (59.4)
GP explains the aim of additional tests	10 (7.5)	29 (21.8)	94 (70.7)
GP explains the results of additional tests	12 (9.0)	33 (24.8)	88 (66.2)
GP explains the method of treatment	6 (4.5)	25 (18.8)	102 (76.7)
GP provides information about the side effects of medication taken	29 (21.8)	39 (29.3)	65 (48.9)
GP provides information about the further course of disease and treatment	11 (8.3)	30 (22.6)	92 (69.2)
GP indicates the date of next visit	6 (4.5)	16 (12.0)	111 (83.5)
GP explains how serious the health problem is	10 (7.5)	31 (23.3)	92 (69.2)
GP asks for concerns about the disease	27 (20.3)	39 (29.3)	67 (50.4)
GP helps the patient to cope with health-related fears and concerns	27 (20.3)	43 (32.3)	63 (47.4)
GP explains how important it is to follow the doctor's instructions	8 (6.0)	26 (19.5)	99 (74.4)
GP is interested in the patient's material situation (provides information about less expensive medicines, the possibility of social assistance)	41 (30.8)	31 (23.3)	61 (45.9)
GP provides information in an understandable way	4 (3.0)	33 (24.8)	96 (72.2)
GP informs about the principles of healthy lifestyle (proper diet, exercise, etc.)	15 (11.3)	41 (30.8)	77 (57.9)
GP proposes preventive examinations (e.g., detection of heart diseases, cancer, etc.)	34 (25.6)	44 (33.1)	55 (41.4)
GP shows interest in the patient's personal situation (specifies a source of social support)	46 (34.6)	34 (25.6)	53 (39.8)
GP is interested in other members of the patient's family (their health, contact among family members)	46 (34.6)	32 (24.1)	55 (41.4)
GP exhibits patience in relation to child patients when they come with parents	21 (15.8)	26 (19.5)	86 (64.7)
GP is taking into account the patient's opinion on the illness of their child	26 (19.5)	33 (24.8)	74 (55.6)
Medical assistance in case of sudden illness	9 (6.8)	19 (14.3)	105 (78.9)
Possibility of talking to the doctor and obtaining clarification by telephone	27 (2.3)	23 (17.3)	83 (62.4)
Doctor related kindly to the patient	3 (2.3)	16 (12.0)	114 (85.7)
Impression that patient's opinion is important for GP	20 (15.0)	29 (21.8)	84 (63.2)
Confidence in the doctor	4 (3.0)	16 (12.0)	113 (85.0)
Receptionist is kind and helpful	9 (6.8)	19 (14.3)	105 (78.9)
Nurses are kind and helpful	4 (3.0)	14 (10.5)	115 (86.5)
Visits are set on the most convenient date	6 (4.5)	22 (16.5)	106 (78.9)

GP general practitioner

problem with obtaining a referral to a specialist ($r = -0.33$, $p < 0.001$) and to additional tests (blood count, RTG, USG) ($r = -0.37$, $p < 0.001$), with a low level of health behavior ($r = 0.28$, $p = 0.003$), including a low level of healthy eating habits ($r = 0.18$, $p = 0.045$), preventive measures ($r = 0.30$, $p < 0.001$), positive attitude ($r = 0.30$, $p < 0.001$), healthy practices

Table 4 Evaluation of satisfaction from healthcare provided by the primary healthcare unit, disease acceptance, quality of life, health behaviors, met needs (group J1, n = 133)

Variables	Mean ± SD	Q25%	Q50%	Q75%	min–max
Satisfaction with primary care services	57.1 ± 13.2	49.0	59.0	68.0	19.0–72.0
Disease acceptance	26.9 ± 7.1	21.0	27.0	32.0	9.0–40.0
Satisfaction with quality of life	3.5 ± 0.9	3.0	4.0	4.0	1.0–5.0
Satisfaction with health state	3.0 ± 0.9	2.0	3.0	4.0	1.0–5.0
Quality of life – physical domain	13.1 ± 2.6	11.4	13.1	14.9	5.1–19.4
Quality of life – psychological domain	12.9 ± 2.5	11.3	13.3	14.7	6.0–19.3
Quality of life – social relationships	14.3 ± 2.7	12.0	14.7	16.0	6.7–20.0
Quality of life – environmental domain	13.5 ± 1.8	12.0	13.5	14.5	8.5–17.5
Health behaviors – total	84.8 ± 14.7	76.0	86.0	95.0	31.0–116.0
Health behaviors – healthy eating habit	3.3 ± 0.8	2.8	3.3	3.8	1.0–5.0
Health behaviors – preventive behavior	3.8 ± 0.8	3.2	3.8	4.3	1.0–5.0
Health behaviors – positive mental attitudes	3.6 ± 0.7	3.2	3.7	4.2	1.7–4.8
Health behaviors – health practices	3.5 ± 0.7	3.0	3.5	4.0	1.5–4.8
Camberwell index (level of met needs)	0.8 ± 0.2	0.7	0.8	0.9	0.2–1.0

Q quartile

($r = 0.20$, $p = 0.033$), with a low level of fulfilled needs ($r = 0.19$, $p = 0.030$), with a low index of benefits ($r = 0.22$, $p = 0.013$).

3.2 Correspondence Analysis

Low levels of satisfaction of patients, both with chronic diseases of the respiratory system (Fig. 1) and with other chronic diseases (Fig. 2), more frequently co-exist with a longer expectation time for an appointment in front of the doctor's office, a long expectation time according to a subjective assessment of patients, difficulties in obtaining a referral to additional tests, a smaller number of appointments at the general practitioner's office within the last 12 months, a higher acceptance of the disease, a lower level of all health behavior (including each category), a shorter duration of the chronic disease, a smaller number of appointments due to chronic diseases, a lower index of benefits, and a smaller number of medicines being taken.

Differences between the study and control groups (Fig. 1 – group J1 and Fig. 2 – group J0) appeared for the following variables: improved well-being in the mental sphere, going to paid doctor's appointments, a problem with obtaining a referral to a specialist and the Camberwell Index.

In the study group, the category "lack of improvement of well-being in the mental sphere" more frequently co-existed with a low level of satisfaction than the category "current improvement of well-being in the mental sphere" with a high level of satisfaction. Also, the category "current improvement of well-being in the mental sphere" more frequently co-existed with a high than low level of satisfaction. In the control group (J0), the situation is reversed: the category "lack of improvement of well-being in the mental sphere" more frequently co-existed with a high than low level of satisfaction, and the category "current improvement of well-being in the mental sphere" co-existed with a low rather than high level of satisfaction. In the study group (J1), improvement of well-being in the mental sphere correlated with an increase in satisfaction, and a deterioration of well-being correlated with a decrease in satisfaction ($r = 0.18$, $p = 0.046$). In the control group (J0), a trend toward the reverse was observed ($r = 0.10$, $p = 0.385$).

A similar result was obtained for the variable "having private appointments". In the study group J1, the category "not having private appointments" more frequently co-existed with a low than high level of satisfaction. The category "having private appointments" more frequently co-existed with a high than low level of satisfaction. In the control group J0, we came

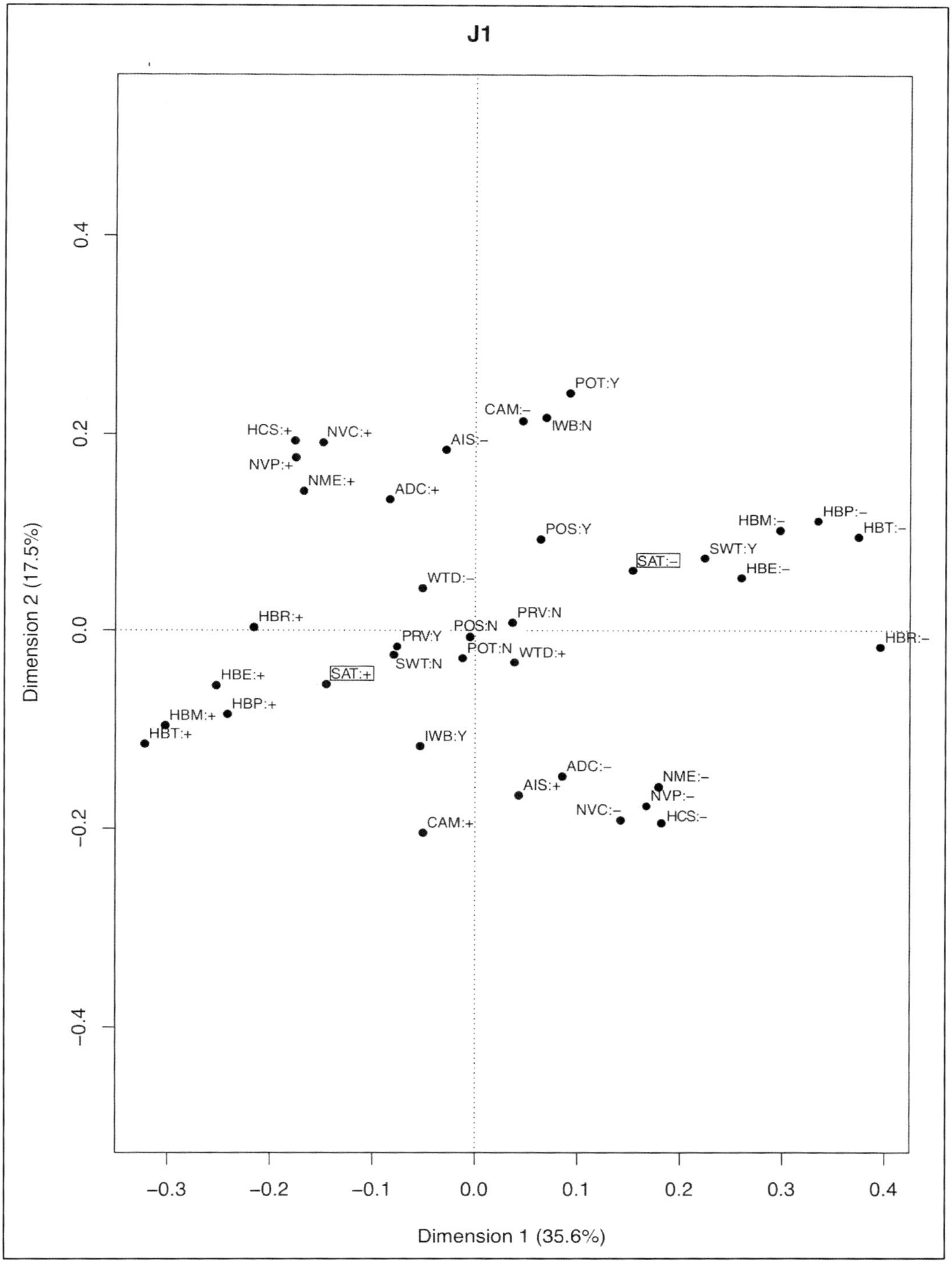

Fig. 1 Correspondence analysis. Response variable is satisfaction (SAT) from healthcare provided by the primary healthcare unit in group J1

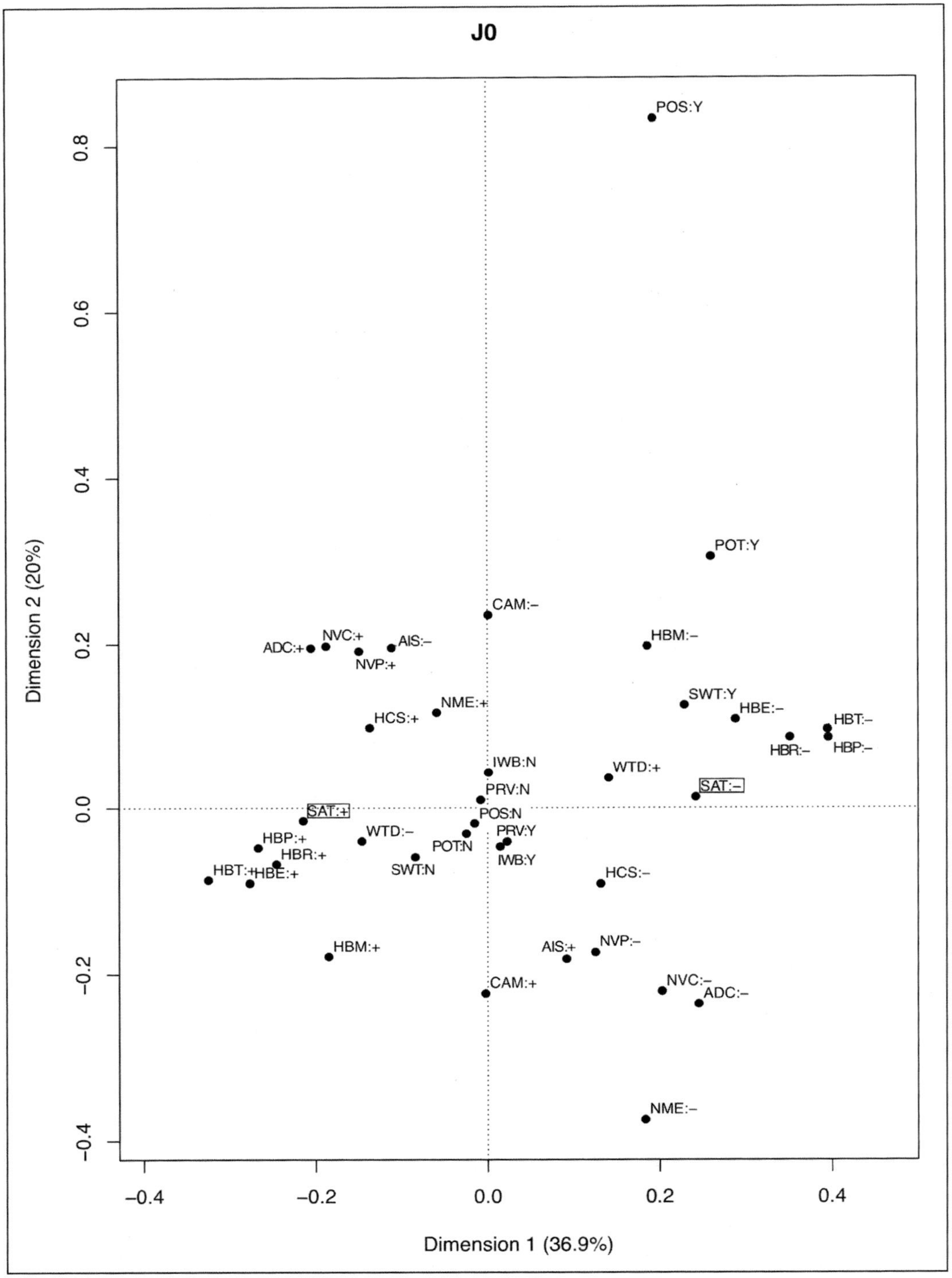

Fig. 2 Correspondence analysis. Response variable is satisfaction (SAT) from healthcare provided by the primary healthcare unit in group J0
IWB improvement of well-being in the past 12 months (IWB:Y | yes, IWB:N | no);

across a reverse situation: the category "not having private appointments" more frequently co-existed with a high than low level of satisfaction and the category "having private appointments" with a low rather than high level of satisfaction. In the study group J1, people who sometimes take advantage of private healthcare were characterized by a higher level of satisfaction than those who do not take advantage of private healthcare, although the correlation coefficient was insignificant ($r = 0.06$, $p = 0.136$). In the control group J0, there was a significant reverse phenomenon: people who sometimes take advantage of private healthcare were less satisfied than those who do not take advantage of private healthcare, with a significant correlation ($r = 0.20$, $p = 0.025$).

In case of the variable "a problem with obtaining a referral to a specialist clinic" there was a difference only for the category "lack of a problem with obtaining a referral to a specialist clinic". In the study group J1, it was difficult to clearly state whether the category more frequently co-existed with a low or high level of satisfaction. In the control group J0, this more frequently co-existed with a high level of satisfaction, with a correlation coefficient of $r = 0.24$ ($p = 0.007$). Therefore, it can be concluded that in the study group J1 the lack of problems with obtaining a referral to a specialist does not increase the level of satisfaction. However, the correlation coefficient calculated for the variables in this group ($r = 0.33$, $p = 0.0001$) indicates that such a dependency can take place.

The last variable for which differences were observed is the Camberwell index. In the study group, the category "a low level of fulfilled needs" more frequently co-existed with a low than high level of satisfaction, and the category "a high level of fulfilled needs" more frequently co-existed with a high than low level of satisfaction. In the control group J0, it was not possible to clearly associate the category "a low/high level of fulfilled needs" with the category "a low/high level of satisfaction". It suggests that in the study group J1 higher levels of fulfilled needs correlated with higher levels of satisfaction. Here, the correlation coefficient $r = 0.19$ ($p = 0.030$) confirms the dependency. On the other hand, in the control group J0 the correlation coefficient was 0.16 ($p = 0.081$), which justifies the lack of a clear assignment of categories described above in this group.

4 Discussion

Defining the level of satisfaction completes the assessment of actual interaction of patients with the healthcare system. Regular and consistent evaluation of the level of satisfaction, with a subsequent program directed at an increase of

Fig. 2 (continued) *WTD* waiting time at the doctor's office (WTD:− | < median, WTD:+ | ≥ median);
SWT subjective evaluation of waiting time (SWT:Y | too long, SWT:N | not too long);
PRV private visits not covered by insurance (PRV:Y | yes, PRV:N | no);
POS problems with obtaining a referral to a specialist (POS:Y | yes, POS:N | no);
POT problems with obtaining referrals for additional tests: blood tests, X-ray, USG (POT:Y | yes, POT:N | no);
NVP number of visits at GP practice (NVP:− | < median, NVP:+ | ≥ median);
AIS level of illness acceptance (AIS:− | < median, AIS:+ | ≥ median);
HBT prohealth behavior – total (HBT:− | < median, HBT:+ | ≥ median);
HBE level of healthy eating habits (HBE:− | < median, HBE:+ | ≥ median);
HBP level of preventive behaviors (HBP:− | < median, HBP:+ | ≥ median);
HBM level of positive mental attitudes (HBM:− | < median, HBM:+ | ≥ median);
HBR level of health practices (HBR:− | < median, HBR:+ | ≥ median);
ADC average duration of chronic disease (ADC:− | < median, ADC:+ | ≥ median);
NVC number of visits due to chronic diseases (NVC:− | < median, NVC:+ | ≥ median);
CAM Camberwell index (CAM:− | < median, CAM:+ | ≥ median);
SAT level of satisfaction from healthcare (SAT:− | < median, SAT:+ | ≥ median);
HCS healthcare services index (HCS:− | < median, HCS:+ | ≥ median);
NME number of medications (NME:− | < median, NME:+ | ≥ median)

quality of care improves the functioning of patients in the clinical and social aspects.

The patient–doctor relationship and healthcare availability are important determinants of quality in the primary care setting (Marcinowicz et al. 2010). The availability of services of general practitioners is a relatively best evaluated dimension in the Polish healthcare system. The weakest part of the system in the social assessment includes difficulties in the availability of specialists and diagnostic tests (CBOS 2014a).

In a Center for Public Opinion Research report (CBOS 2014a), 74 % of respondents stated that it is relatively easy to get an appointment with a general practitioner, patients are treated friendly and with concern (54 % – CBOS report *vs.* 85.7 % – present study), and doctors are engaged in helping patients (58 % – CBOS report). Negative assessments concerned the possibility of getting an appointment for a convenient time (61 % – CBOS report *vs.* 4.5 % – present study), having diagnostic tests quickly (67 % – CBOS report) and getting an appointment with a specialist (85 % – CBOS report) (CBOS 2014a). However, in the present study 8 for 10 patients confirmed that appointments are made on convenient dates.

In one study, the best healthcare system were present in Austria, Germany, and Great Britain, and the worst in Poland (Health Barometer 2012). The Poles assessed the Polish healthcare system at 2.6 on a scale of 1–10 points; the lowest result among all the participating countries: Italy had 3.7, the Czech Republic 4, Sweden 4.7, France, Spain, Great Britain, and Germany had more than 5, USA 5, and Austria 6.5 points, while care for the elderly and the disabled people was assessed in Poland at 2.9 points. Forty one percent of Polish people resigned from medical services due to financial problems (the highest result among all other countries, the lowest of 4 % belonged to Sweden). Of those 20 % resigned from daily healthcare, 17 % resigned from purchasing medicines, 15 % from dental services, and about 10 % from serious treatments or purchasing corrective lenses.

Primary care in Poland was nearly fully (97.0 %) financed by the National Health Fund as of the last quarter of 2013. The situation concerning the ambulatory specialist care was different; every fifth person from the studied population used this kind of care. Only did 63.2 % of them receive benefits from the National Health Fund and 40.2 % financed themselves. The main reason for not going to a specialist was a long waiting time for an appointment (49.8 %) (GUS 2013). According to a CBOS report the main reason for taking advantage of private healthcare is a short waiting time for an appointment (66 %) and more convenient appointment dates (22 %). Younger patients with higher education more frequently use private services, and the rarest use concerns those who frequently ask general practitioners about medical advice (CBOS 2012). The results of our present study indicate that in patients with chronic, other than respiratory, diseases who sometimes use private healthcare, a lower level of satisfaction is observed than in case of people who do not use private healthcare. In patients with chronic diseases of the respiratory system, the lack of problems with obtaining a referral to a specialist has no bearing on the level of satisfaction.

Multiple co-morbidities are common among patients with COPD. Such patients are often prescribed complex medication regimens to be administered by multiple routes for both respiratory and non-respiratory conditions. All these factors predispose patients to the risk of non-adherence to therapy, which is considered the major reason behind emergency hospitalization of COPD patients (Jarab et al. 2012). Adherence to therapy depends on the following key factors: access to a doctor, frequency and duration of appointments, and the quality of the doctor-patient relationship. In the present study, patients received instructions from the GP on the importance of following the course of therapy in a clear majority of cases (74.4 %; Table 3). However, adherence to therapeutic recommendation by Polish patients when treating chronic diseases constitutes a significant problem; numerous patients terminate or modify the recommended pharmacological therapy. Among patients with chronic diseases only 65 % declare that they always take all the recommended doses. Patients in the countries of Western Europe, similar to Poland, also withdraw from the treatment of chronic diseases, they modify dosages or do not start therapy at all. Consequently, it is necessary to use additional medical advice, have diagnostic tests and treatment procedures, including hospitalizations (Polpharma 2010). Among patents with COPD 23.1 % declare that they followed medical recommendations and 33.3 % that they never forget to take the recommended medicines (Napolitano et al. 2015). Nonetheless, non-adherence to medication, especially non-compliance with self-management medication, is very common among asthma and COPD patients (Sari and Osman 2015).

Kardas et al. (2013) indicated that not following medical recommendations by patients has to do with the lack of confidence in doctors and healthcare. In the present study, patients indicated a high degree of confidence in doctors (85 %) (Table 3). A report shows (CBOS 2014b) that 78 % of Poles trust their doctors, but 30 % frequently or always look for information about the diagnosed disease and recommended therapy online, in books or magazines, and 19 % after the consultation with their doctor try to gain a second opinion.

Both effective communication and doctor-patient relationship are key elements in how patients assess their visits to family practitioners (Marcinowicz et al. 2010). Patients with chronic diseases point out the necessity to obtain additional information about the treatment and the most important source of information is a doctor (Napolitano et al. 2015). Patients with COPD feel they are not given enough information about their diagnosis and prognosis, and are keen for more discussion with healthcare professionals. They want more involvement in decisions about treatment and they are glad to discuss general views about future care (MacPherson 2013). From the results obtained within the authors' own studies it can be concluded that only 6 for 10 patients with chronic diseases of the respiratory system had an impression that their opinion is important for the doctor, and that he allows them co-deciding on the type of treatment and further proceedings. The feeling that the patient's opinion is important to the doctor increases the level of co-deciding regarding the diagnostics and therapy by the patient, it also increases the patient's responsibility and changes his approach to health-promoting and proactive activities. Including the patient in the decision-making process results in an increased interest in his health, which is important in the face of not undergoing regular checkups by the majority of Poles (GUS 2012).

A number of patients (38.3 %) participating in the present study were convinced that the doctor knows their medical history (compared with 82.7 % in GUS report) and they were informed by their doctor about the name of disease (82.0 %), causes of disease symptoms (75.2 %), the purpose of additional tests (70.7 %), and the results of tests (66.2 %). Moreover, the doctor informed patients about the further course and treatment of disease (69.2 % compared with 23.3 % in GUS report) and explained the seriousness of a health problem (69.2 %). What is important, the information was provided in a comprehensible way (72.2 % compared with 33.3 % in GUS report (GUS 2012), and patients were treated in a friendly manner and with concern (85.7 %). These data indicate a high level of responsiveness of primary care doctors toward patients.

The patients were informed about side effects of the medicines being taken to a different extent (48.9 % – yes, 21.8 % – no, 29.3 % – sometimes). It is worth paying attention to the fact that if the doctor excessively highlights the dramatic consequences of a disease in cases when patients have chronic diseases, it frequently causes a reverse reaction from the patient, increases fear, and causes intensification of mechanisms of repression and denial (Polpharma 2010).

The results of our studies demonstrate that in patients with chronic diseases of the respiratory system improvement of well-being in the mental sphere has to do with an increase in satisfaction, and deterioration of well-being with a decrease in it. The lack of problems with obtaining a referral to a specialist and a higher level of fulfilled needs have a similar influence. Other authors state that the occurrence of a chronic disease itself positively increases the level of satisfaction (Carlin et al. 2012), higher levels are also reported in women and older patients with a lower level of education, residents of smaller towns, who have doctor's appointments more often and less intensive clinical symptoms (Petek et al. 2011).

The majority of patients in the present study did assess negatively, as the lack of interest or sporadically expressed interest, of the GP in the patient's material and personal situation (specifically in the source of social support; Table 3). Eisner et al. (2011) found that socio-economic status represents a risk factor for adverse COPD health outcomes. Low education levels and household income were consistently related to greater disease severity, poorer lung function, and a greater physical functional. On the other hand, results of a different study (Trachtenberg et al. 2014) indicate that patients with COPD, who have a lower income were more likely to be hospitalized than peers having the highest income. These authors also suggests a necessity for a broader look at socioeconomic health determinants from the perspective of a reduction of the number of hospitalized patients, which could lead to considerable reductions in healthcare costs (Trachtenberg et al. 2014).

Indicators of the care continuity, i.e., the possibility of having an appointment and good relationship with the same doctor, and a sufficient consultation time influence the appointment quality most from the patient's point of view. In the present study, the time committed by the doctor was sufficient for 82.7 % of patients (*vs.* 63.5 % of patients in a GUS report) (GUS 2012), and 72.9 % of patients confirmed that the time was sufficient for a conversation.

Patients indicating higher levels of satisfaction provide their doctor with more information during appointments, they ask more questions, remember more information obtained from the doctor during appointments, and follow the doctor's recommendations related to the therapy plan more accurately. Patients with a positive approach to the appointment and prepared by obtaining information from different sources are provided with more services related to prevention (Tarn et al. 2012). Therefore, the assessment of the level of satisfaction is essential for the process of promoting health and prophylactics of diseases as basic elements of modern primary care.

5 Conclusions

The level of satisfaction regarding healthcare is related to the possibility to co-decide about the course of therapy and prevention. Obtaining information by patients and the possibility to ask questions influence the improvement of satisfaction from the quality of care. The present findings indicate that in patients with chronic diseases of the respiratory system the increase in satisfaction depends on improvement of well-being in the mental sphere. The lack of problems with obtaining a referral to a specialists and a higher level of fulfilled needs have a similar influence. Additionally, low levels of satisfaction should be expected in patients with chronic diseases of the respiratory system, who are waiting for an appointment in front of the office for a long time, report their problem with obtaining a referral to additional tests, with a low level of health behavior, and with a low index of benefits.

Conflicts of Interest The authors declare no conflicts of interest in relation to this article

References

Carayon P, Hundt AS, Hoonakker P, Kianfar S, Alyousef B, Salek D, Cartmill R, Walker JM, Tomcavage J (2015) Perceived impact of care managers' work on patient and clinician outcomes. Eur J Person Centered Healthc 3(2):158–167

Carlin CS, Christianson JB, Finch M (2012) Chronic illness and patient satisfaction. Health Serv Res J 47 (6):2250–2272

CBOS – Center for Public Opinion Research (2012) Poles about national and private healthcare. http://www.cbos.pl/SPISKOM.POL/2012/K_047_12. Accessed 15 Dec 2012

CBOS – Center for Public Opinion Research (2014a) Statement regarding the studies conducted by CBOS. Opinions about the functioning of the healthcare system. Nr 107/2014, Warszawa 2014. http://www.cbos.pl/SPISKOM.POL/2014/K_107_14.PDF. Accessed 17 Aug 2015

CBOS – Center for Public Opinion Research (2014b) Statement regarding the studies conducted by CBOS. Opinions about medical errors and trust in doctors. Nr 165/2014. Warszawa 2014. http://www.cbos.pl/SPISKOM.POL/2014/K_165_14.PDF. Accessed 18 Aug 2015

Eisner MD, Blanc PD, Omachi TA, Yelin EH, Sidney S, Katz PP, Ackerson LM, Sanchez G, Tolstykh I, Iribarren C (2011) Socioeconomic status, race and COPD health outcomes. J Epidemiol Community Health 65(1):26–34

GUS – Central Statistical Office of Poland (2012) Do doctors meet patients' expectations? http://www.mp.pl/kurier/65386. Accessed 18 Aug 2015

GUS Central Statistical Office of Poland (2013) Health and healthcare 2013. http://stat.gov.pl/obszary-tematyczne/zdrowie/zdrowie/zdrowie-i-ochrona-zdrowia-w-2013-r-,1,4.html. Accessed 18 July 2015

Health Barometer 2012 Europe Assistance Health Barometer: Healthcare in Europe and in the USA (2012). Available from http://www.europ-assistance.com/sites/default/files/2012_csa-europ_assistance_health_barometer-global_report.pdf. Accessed 20 Aug 2015

Jarab AS, AlQudah SG, Khdour M, Shamssain M, Mukattash TL (2012) Impact of pharmaceutical care on health outcomes in patients with COPD. Int J Clin Pharm 34(1):53–62

Jaracz K, Kalfoss M, Górna K, Baczyk G (2006) Quality of life in Polish respondents: psychometric properties of the Polish WHOQOL-Bref. Scand J Caring Sci 20:251–260

Juczyński Z (2001) Assessment tools in health promotion and health psychology. Polish Psychological Association, Psychological Tests Laboratory, Warsaw

Kardas P, Lewek P, Matyjaszczyk M (2013) Determinants of patient adherence: a review of systematic reviews. Front Pharmacol 4:91. doi:10.3389/fphar.2013.00091

MacPherson A (2013) The views of patients with severe chronic obstructive pulmonary disease on advance care planning: a qualitative study. Palliat Med 27 (3):265–272

Marcinowicz L, Rybaczuk M, Grebowski R, Chlabicz S (2010) A short questionnaire for measuring the quality of patient visits to family practices. Int J Qual Healthcare 22(4):294–301

McKee M, Jakab Z, Andriukaitis V, Barnhoorn F (2015) European Public Health News. Message from the WHO regional director for EuropeHow eHealth can help with Europe's chronic diseases epidemic8th EPH conference – 'Health in Europe – from global to local policies, methods and practices. Eur J Pub Health 25:748–750

Napolitano S, Napolitano P, Angelillo IF, Collaborative Working Group (2015) Medication adherence among patients with chronic conditions in Italy. Eur J Public Health. http://eurpub.oxfordjournals.org/content/early/2015/08/11/eurpub.ckv147. Accessed 18 Aug 2015

Petek D, Künzi B, Kersnik J, Szecsenyi J, Wensing M (2011) Patients' evaluations of European general practice-revisited after 11 years. Int J Qual Healthcare 23:621–628

Polpharma – Polish patient's self-portrait (2010) A report about the adherence to therapeutic recommendations by Polish patients. Patients' approach to medical recommendations in the therapy of chronic diseases. Polpharma. Warsaw, April 2010. http://www.polpharma.pl/gfx/polpharma/pl/polpharmakatalogfleshowy/334/2/1/raport_z_badania_screen.pdf. Accessed 17 Aug 2015

Pourat N, Davis AC, Chen X, Vrungos S, Kominski GF (2015) In California, primary care continuity was associated with reduced emergency department use and fewer hospitalizations. Health Aff (Millwood) 34 (7):1113–1120

Quigley DD, Elliott MN, Farley DO, Burkhart Q, Skootsky SA, Hays RD (2013) Specialties differ in which aspects of doctor communication predict overall physician ratings. J Gen Intern Med 29(3):447–454

Sari N, Osman M (2015) The effects of patient education programs on medication use among asthma and COPD patients: a propensity score matching with a difference-in-difference regression approach. BMC Health Serv Res 15:332

Sebo P, Herrmann FR, Bovier P, Haller DM (2015) What are patients expectations about the organization of their primary care physicians'practices? BMC Health Serv Res 15:328

Smidth M, Olesen F, Fenger-Grøn M, Vedsted P (2013) Patient-experienced effect of an active implementation of a disease management programme for COPD – a randomised trial. BMC Fam Pract 14:147

Sultan N, Khuwaja AK, Kausar S, Nanji K (2012) Patients' evaluations of family practice care and attributes of a good family physician. Qual Prim Care 20:375–383

Tarn DM, Young HN, Craig BM (2012) Development of the patient approach and views toward healthcare communication (PAV-COM) measure among older adults. BMC Health Serv Res 12:289

Trachtenberg AJ, Dik N, Chateau D, Katz A (2014) Inequities in ambulatory care and the relationship between socioeconomic status and respiratory hospitalizations: a population-based study of a Canadian city. Ann Fam Med 12(5):402–407

Advs Exp. Medicine, Biology - Neuroscience and Respiration (2016) 21: 87–88
DOI 10.1007/978-3-319-30659-9

Index

Printed by Printforce, the Netherlands